INTRODUCING GEOLOGY

INTRODUCING GEOLOGY

The earth's crust considered as history

by
D. V. AGER

FABER AND FABER
3 Queen Square
London

First published in 1961
by Faber and Faber Limited
3 Queen Square London WC1
First published in this edition 1966
Reprinted 1968
Second edition 1975
Printed in Great Britain by
Latimer Trend & Company Ltd Plymouth
All rights reserved

© D. V. Ager
1961, 1975

ISBN 0 571 04858 7 (*Faber Paperbacks*)
ISBN 0 571 04857 9 (*hardbound edition*)

To my Mother and Father

CONTENTS

ILLUSTRATIONS

LIST OF PLATES
after page 32

LIST OF TEXT FIGURES

9

ILLUSTRATIONS

ILLUSTRATIONS

PREFACE TO THE SECOND EDITION

When I wrote the first edition of this book in 1958, I commented that 'a geologist will undoubtedly be among the first men on the moon'. Since then we have all seen the strange white figures bouncing around on the lunar surface and heard distorted American voices using geological terms such as 'phenocryst'. And before the end of this year (1973), the first geologist on the moon, Dr. Harrison Schmitt, will be talking to the Geological Society of London.

However, it is on planet Earth that the most startling geological discoveries have been made in the last few years. In 1958, the terms 'sea-floor spreading', 'plate tectonics', 'transform fault' and 'subduction zones' had never been heard in geological circles (though some of them had been implied by that great man Arthur Holmes). These ideas have produced what has quite properly been called a 'revolution in the earth sciences'.

Revising this book in the light of these new ideas has been a difficult task, since my original intention (which I have retained) was to present geology in the light of the British geological record, which was likely to be both the most familiar and the most accessible to British readers. But the theories of plate tectonics need a broader and a more theoretical approach. The result is a compromise. However, I sometimes think that the more passionate preachers on plate tectonics tend to forget that, though these concepts have undoubtedly revolutionized our ideas on earth history, there is still a great deal of exciting geology to be studied and unravelled for which we don't need the standard bent escalator diagram of a Benioff Zone. That kind of diagram (e.g. Figure 5 in this book) even appears now surrounded by Chinese characters and one almost welcomes a paper without one.

Also if I were writing this book for the first time today, when

13

foreign travel has become so commonplace, I would probably not adopt such an insular approach as I did in 1958. But the geology is still there in Britain and it's still just as fascinating, even if it is looking a little more battered now from generations of passing enthusiasts.

Swansea
November, 1973   D. V. AGER

FOREWORD

This book is not an 'Introduction to Geology', which would be a standard elementary textbook dealing solemnly (but inadequately) with all aspects of the subject. There are plenty of textbooks already. This is called *Introducing Geology* and is intended to do just that—in the best way possible—through the medium of the familiar. This book concentrates mainly on Britain, not through insular narrow-mindedness, but because we have such an unusually complete and varied rock record in these islands, and because the places mentioned are likely to be known to the reader. Nevertheless, other parts of the world are discussed when they illustrate points or theories missing or not clearly expressed in our own record.

It is not my concern to give a detailed account of the geologist's methods, or even to consider precise evidence, but merely to summarize the story that has emerged in more than 150 years of scientific study of geological history. Readers will find only the minimum of technical terms, and no fossils or minerals are named if it can possibly be avoided. This book is intended not so much to instruct as to interest. The various aspects of the subject, the various methods, the various types of evidence, all are included where they fit naturally into the story.

Geology is simply history extended backwards for millions of years. It is not really a science at all, but rather the application of almost every other science to the elucidation of earth-history. Just as a modern archaeologist uses isotope chemistry to date his ancient documents and aerial photography to find his buried walls, so the geologist uses almost every branch of chemistry, physics and biology to investigate his rocks and their contents. Apart from the strictly economic aspects of the subject—such as finding ore bodies or underground water supplies—the geologist is primarily concerned with working out the past history of the earth. All branches of the subject are essentially contributory to this central theme. This book therefore has the sub-title 'The Earth's Crust Considered as History'.

Since I am not concerned with the horrors of a syllabus or of examinations, I have allowed myself the luxury of a bias towards the 'rocks with some life in them' and towards organic evolution generally. I do not apologize for this, it reflects my own interests, and in my experience, the interests of the majority of general readers. In the Appendix there are a number of suggestions for further reading and other activities. If this book serves its purpose, then it is hoped that many who have not met geology before will read more about the subject, and better still, they will go and look at the rocks for themselves.

Urbana, Illinois
November, 1958

ACKNOWLEDGEMENTS

A conscientious scientist who writes a book like this must feel that his chief debt is to the contemporaries and predecessors who actually did the work and made the discoveries of which this is such a meagre record. The generalizer and the popularizer are usually regarded somewhat doubtfully by the workers of science, and often rightly so, but the author asks forgiveness for the obvious scientific sins contained in this work.

Apart from Plates 20 and 21, which were taken by the author, the plates all come from the excellent photographic collection of the Institute of Geological Sciences at the Geological Museum in South Kensington. The writer acknowledges with thanks the permission given by the Controller of Her Majesty's Stationery Office for these to be reproduced. He also allowed the use of certain published Survey figures as bases for some of the drawings in this book.

The final versions of the text figures were drawn by Mr. Walter Shepherd and Mr. J. B. Frame whose co-operation, understanding and ability were of considerable assistance to the author. Miss Elaine Bryant, quite apart from her full-time duties as research assistant, helped a great deal with the preparation of the original diagrams. She also read and usefully criticized a large part of the text. Her successor, Miss Mary Pugh, helped considerably with the checking of proofs and the preparation of the index. The new figures (and revised old figures) for the Second Edition were drawn by John Uzzell Edwards. Miss M. Davies and Mrs. K. Guntrip kindly typed the revised text.

Finally the author must record his very sincere thanks to his wife for her assistance, encouragement, constructive criticism and extreme tolerance during the preparation of this book.

I

INTRODUCING GEOLOGY

The geologist studies the earth as it is today and as it has been throughout its long history. He is interested in every aspect of the history of the earth, its changing geography, its life, its climate, the way the frost breaks away the tops of the highest mountains and the way mud accumulates in the deepest parts of the sea. Being mere man, the geologist can only study the surface of his planet. His deepest mines are but the faintest pinpricks in the outermost tissue of the globe. By using the methods of modern physics, the geologist can make some inspired guesses as to what lies below but his first concern is with rocks at the surface and with the natural processes which affect them.

There are two main reasons why geologists study the rocks. They may do it in response to the scientist's natural and insatiable curiosity about his world, or they may do it in search of the raw materials essential to man's present mode of life. But there is no sharp division between the pure and the applied sides of geology.

For more than a hundred years the geologist was always among the pioneers who first ventured into unknown territory. He often arrived before the missionary, the trader and the government administrator. At this moment geologists are actively engaged in exploring the remoter parts of the world, ranging from the icy mountains of Alaska to the scorching heat of the Sahara. They are also very much occupied in the expensive business of exploring the sea and ocean floors. More and more papers are published every year on the geology of the moon and Mars. Many of such geologists are employed by large oil companies and mining syndicates, and are primarily concerned with finding valuable mineral deposits; others are working for government surveys or on government-financed expeditions. Whoever employs them, the first duty of the field geologist is to produce a map; he has to record what is there, whether or not it is of immediate material use. As a territory becomes better known, the geological survey becomes more and more detailed and precise, and the geologist uses the discoveries of many other branches of science to help him. As the likelihood of large and spectacular finds of valuable minerals diminishes, so the known occurrences are more fully explored on the surface and underground. This work of geological exploration and map-making shows no signs of being completed, even over the most-studied parts of the earth's surface. In Britain, where the Institute of Geological Sciences (formerly called the Geological Survey) has been mapping for nearly 140 years, the work is far from complete and some 50 university and polytechnic departments still find plenty of original work for their students and researchers.

The purely economic aspects of geology fall into several categories. Firstly there is the search for useful metallic ores such as iron, lead, tin and zinc, not forgetting rarer and more valuable ores such as gold and, nowadays, uranium. There are also many non-metallic minerals with which the geologist is concerned, for example sulphur and rock-salt.

Britain's industrial greatness was built on her thick and extensive deposits of coal, and geologists, especially those of the government survey, are much concerned with the exploration of every possible underground reserve of this valuable fossil fuel. The wealth of many other countries depends almost entirely on their hidden resources of another fossil fuel—oil—and a very large number of geologists, specializing in different fields, are employed in the oil industry.

Underground water is another valuable commodity which comes very much into the geologist's province. The finding of suitable supplies is far better placed in the scientific hands of a geologist with his maps than in the superstitious hands of a diviner with his hazel twig.

All large civil engineering firms now either employ or consult geologists in connection with major constructional works such as dams, airport runways, tunnels and bridges. Many thousands of pounds have been wasted through heavy structures being sited in places which were geologically unsuitable. Geologists advise on such matters as the nature of the rocks likely to be encountered underground, their probable strength under load and the suitable rocks likely to be available for building stone or for making concrete.

Apart from, and yet underlying all these applied aspects of geology, there is the central theoretical basis of the subject which is the chief theme of this book. This arises directly from the work of the geological explorer and map-maker. As the rocks of an area are studied and correlated, the fossils collected and the complicated structures unravelled, there emerges a coherent story of the history of that area extending back hundreds of millions of years. This is the great fascination of the subject. It is all one long, exciting and awe-inspiring detective story. New clues are always being found, in the field, in the laboratory and in the writings of other geologists. Slowly the isolated items of information fit together and a whole chapter of the story suddenly becomes plain. It is a story with a vastly greater time-scale than history or archaeology, but still dealing with solid things one can handle, unlike the unapproachable mysteries of astronomy. The story covers hundreds and thousands of millions of years and deals with the evolution of our planet and the life it bears, from its fiery birth to what it is today.

The one fundamental principle which underlies all geological thought is that the earth is always changing. The mountains are being raised up and worn down, the seas are advancing and retreating, rocks are being altered by heat and pressure, molten rock is being pushed through the crust and bursting out at the surface as volcanoes, animals and plants are changing from one species into another. All this seems to have been going on since the first formation of the earth's crust and the first appearance of life, and all, it seems, as a result of the same natural processes which are operating today. This is a very different picture from that of the early thinkers who accepted the world as it is, and thought of it as having been thus since the beginning of time. It is also different from the ideas of later natural philosophers who postulated great periodic cataclysms. The form of our present-day world is a fleeting, ephemeral thing, but at the same time there are very few phenomena of the past, recorded in our rocks, which we cannot see taking place somewhere today. The anatomy is always changing, but the physiology always remains the same.

Geology came into its own as a science when thinkers started going to look at the rocks for themselves instead of merely theorizing about them in their studies. The armchair thinkers are still with us, but the reader of this book would be best advised to leave his armchair as soon as possible and to get out into the field to see for himself.

THE BRANCHES OF GEOLOGY

Like all modern sciences, geology has become divided into a number of branches, and very few geologists can be specialists in more than one branch, or real experts in more than one tiny aspect of one branch. Even so, there are many geologists who pride themselves on being 'field men', experts in no particular branch of the subject except in that of observing and recording what they see out of doors.

The different 'specializations' may usefully be considered in turn.

Historical geology (stratigraphy): this is the heart of the matter and belongs to no other subject but geology. It is the story which has been worked out from the evidence collected by all the other branches. It is the main concern of this book.

The study of rocks and minerals (petrology and **mineralogy):** this is the chemistry and physics of geology. It includes the identification of minerals and the study of the nature and possible origin of the many different types of rocks. It should be noted here that 'rock' to a geologist means anything from a hard granite to a soft unconsolidated mud.

The study of fossils (palaeontology): this is the biological side of geology. Fossils are the remains of past life or direct evidence of the existence of that life (for example, footprints). The study of fossil plants is usually separated off as **palaeobotany**, and that of fossil animals is sometimes called **palaeozoology**. Fossils are interesting on their own account as records of past life and its evolution. They are also essential to geologists in the dating and correlation of rocks.

The study of rock structure (structural geology or **tectonics):** this is the branch of geology which is concerned not so much with the rocks themselves as with the ways in which they have been deformed, folded and broken since their formation. It is the mechanics of geology. The structural geologist studies the disposition of the rocks at the earth's surface at the present day, and how they came to occupy their present positions.

The study of landscape and erosion (physical geology or **geomorphology):** this is often regarded by geologists as a branch of that indefinable subject geography, but it is in fact inseparable from the other branches of geology. The present landscape bears many marks which reveal its past geological history, and the rate and mode of erosion is closely controlled by the nature and structure of the rocks.

Geophysics: this is on the borderline between geology and physics. It concerns the investigation of the earth, especially its interior, by various physical methods, such as the accurate measurement of gravity, magnetism, electrical resistivity and seismic waves.

Geochemistry: this is the study of the chemical composition of the earth's crust and includes the dating of rocks by means of their contained radio-active elements. Other studies include the tracing of ore bodies by means of the chemical elements found in the adjacent soil or plants.

Engineering Geology: this is the application of geology to engineering problems. It is particularly concerned with large structures such as bridges, dams and tunnels. It also concerns the study of underground water supplies.

THE DIVISIONS OF GEOLOGICAL TIME

We think of human history in terms of centuries, decades and years. Such time divisions are necessary so that we can organize our thoughts and can correlate the past happenings in different fields of activity and in different places. In the same way, subdivisions are needed for the vastly longer period of earth history, and for this purpose hundreds, thousands or even millions of years are units too small and too difficult to distinguish. Instead there has grown up a set of major divisions which are used all over the world, but which are founded on the European rock succession and have not yet been tied on to an exact time-scale. There are also many smaller and smaller subdivisions which are progressively less widespread in application.

On the largest scale there is the two-fold division of geological time into an inconceivably long period before the appearance of obvious life, and a much shorter period during which life has been present. All the rocks which were formed on the earth's surface (or beneath it) before the appearance of life are generally known as **Precambrian**. There are names for many subdivisions of this in various parts of the world, but there is, as yet, no sure way of correlating these ancient rocks from place to place. The later rocks are much easier to handle, for in them there are the fossilized remains of past faunas and floras, and with these fossils we can judge the age of the rocks which contain them.

The fossiliferous rocks are divided into a number of **Systems**, starting with the Cambrian at the bottom and finishing with the Holocene or 'Recent' at the top (which includes the rocks still being formed at the present day). A 'system' is essentially a set of rocks, and these were formed during a length of geological time which we call a **Period.** Thus the Devonian System is the great heap of rocks which accumulated during the Devonian Period. These periods are given in the table on page 25, together with their (very approximate) age in years.

The geological periods are grouped into four **Eras** (also shown on the table): the Palaeozoic (or Primary) Era which was very long, the Mesozoic (or Secondary) Era, the Tertiary Era and the Quaternary Era. These last two are often put together as the Caenozoic Era. The names given in the table are used all over the world, though the Americans have two periods—the Mississippian below and Pennsylvanian above in place of the Carboniferous. Many geologists also still follow the old custom of regarding the Eocene, Oligocene, Miocene and Pliocene as four distinct (though very short) periods within the Tertiary, and many add a further division the Palaeocene—at the beginning of that era.

It will be seen that a geological period has no standard length. Nevertheless, the scheme which grew up in Britain and western Europe generally, proved to be soundly based and can now be applied almost everywhere. Above the Precambrian, the divisions are founded on major changes in the organic population of the earth's surface, which seem to have occurred more or less simultaneously all over the world. This approach has many snags and limitations, some of which will be discussed later, but they do not affect the general pattern of geological evolution.

THE MAIN TYPES OF ROCK

It is the intention in this book to introduce the various terms and principles of geology as they arise naturally in the record. Thus fossils will first be discussed in Chapter III and the folding of rocks in Chapter IV. Explanations of special terms can be found in the glossary at the end of the book. It is desirable, however, to

THE STRATIGRAPHICAL COLUMN

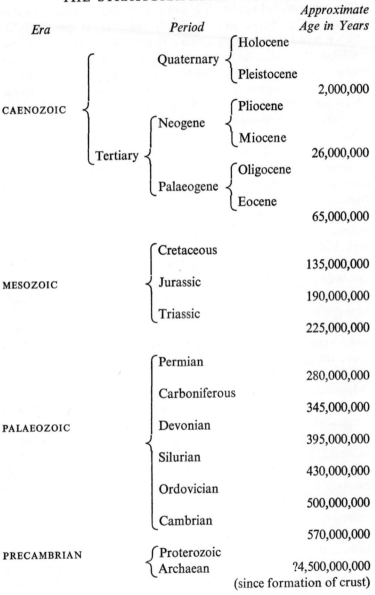

Era	Period			Approximate Age in Years
CAENOZOIC	Quaternary		Holocene	
			Pleistocene	
				2,000,000
	Tertiary	Neogene	Pliocene	
			Miocene	
				26,000,000
		Palaeogene	Oligocene	
			Eocene	
				65,000,000
MESOZOIC	Cretaceous			
				135,000,000
	Jurassic			
				190,000,000
	Triassic			
				225,000,000
PALAEOZOIC	Permian			
				280,000,000
	Carboniferous			
				345,000,000
	Devonian			
				395,000,000
	Silurian			
				430,000,000
	Ordovician			
				500,000,000
	Cambrian			
				570,000,000
PRECAMBRIAN	Proterozoic			
	Archaean			?4,500,000,000
				(since formation of crust)

say a little about the main categories of rock at this stage, rather than to try to explain them in the unsuitable setting of the Precambrian.

The conventional approach is to say that there are three fundamentally different types of rock. Firstly there are the **igneous rocks,** which reached their present position in a molten state and are now solidified. Secondly there are **sedimentary rocks,** which are the accumulated debris produced by the wearing down of older rocks by many different agencies. Finally there are the **metamorphic rocks,** which are either igneous or sedimentary in origin, but which have been altered, sometimes out of recognition, by heat or pressure.

The greater part of this history concerns the sedimentary rocks which accumulated very, very slowly, layer upon layer. The most common sedimentary rocks are those which were originally laid down on the sea floor. They consist of the material brought down by rivers, or by glaciers, or blown out to sea by the wind, or broken off from the cliffs by the sea itself. This debris is associated with the remains of sea animals, which sometimes form the major part of the rock. Sedimentary rocks may also accumulate on the flood-plains of rivers, in lakes or actually on dry land—as in deserts. By being laid one bed upon another in regular succession, they retain for us a more or less detailed record of the happenings in a particular area through long periods of geological time. Thus, providing there have been no great disturbances of the crust in the area, the higher layers of rock will be younger and the lower ones older. This 'Law of Superposition', though seemingly obvious, was not recognized by natural philosophers for hundreds of years.

Igneous rocks on the other hand, being more or less liquid when they are put in place, usually do not observe this law. They may be erupted at the surface as volcanic ash or lava flows, which cool and harden. In this case they take their place in the record like a bed of sediment. But to reach the surface from the great depths where they originate, they must push their way through many layers of sedimentary rock. They may cool and solidify in this position. Thus igneous rocks of a particular age may be found cutting through any sediments which are older than themselves.

Igneous rocks which were originally erupted at the surface are said to be **extrusive.** Those which solidified below the surface are said to be **intrusive.** Intrusive igneous rocks may take various forms. They may force themselves along the bedding of the sediments as horizontal sheets or **sills.** They may cut through the bedding as more or

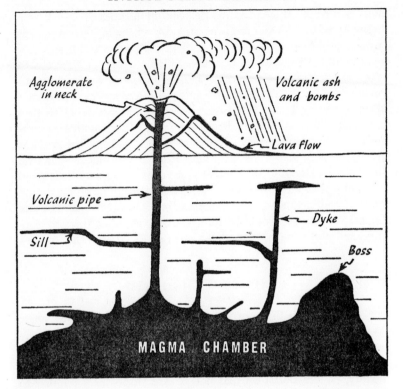

FIG. 1

Forms of igneous intrusions and extrusions.

less vertical sheets or **dykes**. They may form narrow **pipes** leading to the crater of a volcano, or they may occur as great irregular masses. Some of the forms shown by extrusive and intrusive igneous rocks are illustrated in Figure 1.

When igneous rocks are intruded in a molten state, they are usually very hot, and they may affect the rocks about them in various ways. This is one way in which the third category of rocks—the metamorphic rocks—may be produced. There are often small areas of metamorphosed sediment around large igneous intrusions. The sediments may be baked, with new minerals formed by chemical changes produced at high temperatures. But these are merely local effects, and metamorphic rocks cover great areas of the earth's surface. Most of them were produced by what is called 'regional

27

metamorphism'. Huge areas and tremendous thicknesses of rock have been altered by heat and pressure, probably when they were buried at great depths beneath the earth's surface. Only great uplift and very long periods of erosion have exposed these metamorphic rocks to our view at the present day. They may represent the roots of ancient mountain chains which have long since been worn away. The deeper these rocks are buried, the higher the temperatures to which they are subjected, and at great depths the roots of the mountains must melt and flow like treacle. They are then, by definition, igneous rocks, and may solidify as such. So it is often very difficult to draw a line between igneous and metamorphic rocks. Professor H. H. Read has pointed out that our conventional classification of rocks into igneous, sedimentary and metamorphic is unsatisfactory, and a more logical subdivision would be into volcanic, neptunic (i.e. sedimentary) and plutonic (i.e. formed at a great depth beneath the surface).

THE NEW GEOLOGY

Since the first edition of this book was published in 1961, there has been a major revolution in the science of geology and we are now in the second exciting 'heroic age' of the subject. Though this book still maintains the approach of presenting geology as earth history, concentrating on that part of the world most familiar to British readers, it is now inappropriate to plunge straight into that subject without first considering some of the major processes that seem to have controlled the record of the rocks. We are now in the happy position of having—for the first time—a general theory of geology to which everything else must be related. In the early days of geology, the first clues to understanding the earth were provided by the doctrine of **uniformitarianism**. This stated simply that the rocks can only be understood by relating them to the processes seen going on at the present day. We can recognize that certain ashy-looking rocks were blasted out of volcanoes millions of years ago by comparing them with the products of modern volcanoes. We can interpret ancient sandstones as deltaic deposits by comparing them with the sediments laid down in present-day deltas. This doctrine, which is discussed again in Chapter XII, is expressed most clearly in the epigram 'The Present is the Key to the Past'. There are limitations to this way of reasoning, but it has provided the basic philosophy which has enabled us to work out the geological history of our planet.

Now, in the 'new' geology we still use a uniformitarian approach, but we apply it not only to the surface phenomena such as sedimentation and erosion, but also to the internal processes we can detect going on inside the earth.

The new 'general theory' is that of **plate tectonics**, which postulates that the crust of the earth consists of a number of rocky plates, carrying continents and oceans, which move about relative to one another as a result of the internal forces (presumably convection currents) within the earth.

It has long been suspected that the continents have moved about on the earth's surface through geological time. This is suggested by the way coastlines, for example those on either side of the South Atlantic, match one another, not only in outline but also in geology. This will be discussed further in Chapter VII. Strong support for this theory of continental drift, as it was called, was provided by the geophysicists when they found that many rocks carry what is called 'residual magnetism', indicating the position of the poles at the time they were formed. It is as though each continent carried a large number of compasses, each locked in position at some time in the past. Certain minerals oriented themselves in line with the magnetic field of the earth, and if the continent that carries them has moved since they were formed, then obviously we have a record of the position of the relevant continent through geological times. At least, it is comparatively easy to deduce latitude this way (by the angle of inclination of the magnetic field) but it is not possible to deduce longitude. Another surprising fact discovered by the geomagnetists is that the earth's magnetic field has frequently, and apparently quite suddenly, reversed itself in the past.

'Plate tectonics' is not just another term for 'continental drift', since the earlier theory just had the continents moving, whereas now we think of plates moving which carry on their backs either continents or oceans (which are fundamentally different in composition). An essential corollary of plate tectonics and the idea that got the whole thing going, was the theory of sea-floor spreading.

The basic idea is that as two plates move apart, a split develops between them up which comes molten material from the earth's interior. We can see this happening most clearly in the middle of the Atlantic, where the ocean floor is tearing along a dotted line down its centre. This line is marked by volcanic islands extending from Iceland in the north, down via the Azores and a whole string of smaller islands to lonely Tristan da Cunha in the far south. Only this year (1973) a new volcano appeared in the island of Heimaey, south of Iceland, as the split opened once again and more molten material came from the earth's interior to solidify as new ocean crust.

The split down the centre of the Atlantic can be traced all the way (Figure 2) as a ridge of volcanic rock (the Mid-Atlantic Ridge). As the split continued to develop through the countless millenia of geological time, the continents on either side inevitably grew farther apart. In other words, the two sides of the ocean, which once fitted so snugly together, were separated by an ever-widening slab of oceanic crust. That the process happened in this way can be demonstrated quite clearly by the age of the rocks concerned. Thus the volcanic rocks in the centre of Iceland are of Quaternary age, whereas those in the east and west of that island are of the late Tertiary. What is more, if one continues to our side of the Atlantic, one finds in the Inner Hebrides and on the west coast of the Scottish mainland, volcanic rocks of early Tertiary or even greater age.

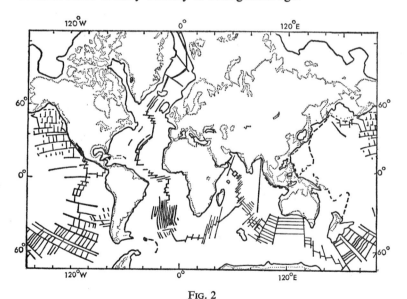

FIG. 2

The mid-oceanic ridges of the world along which sea-floor spreading is taking place at the moment. The continents are defined by the edges of the continental shelves. The transverse lines represent transform faults.

Another line of evidence is provided by the magnetism of the volcanic rocks. As they cooled from the molten state they preserved in the orientation of certain of their minerals the direction of the earth's magnetic field at the time. As already explained, we now know that this frequently changed. So the rocks of the Atlantic floor show the 'north' pole first in the north, then in the south, flipping backwards and forwards in the magnetic reversals mentioned above. So we get what is known as magnetic 'striping' on the sea-floor which shows a remarkable mirror image of itself from one side of the Atlantic to the other. Figure 3 shows how these magnetic stripes

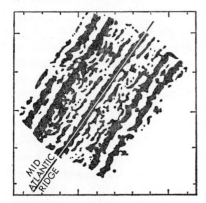

Fig. 3
Diagram showing 'normal' (black) and 'reversed' (white) magnetism of the sea-floor south-west of Iceland.

match on either side of the ridge running south-west from Iceland. The extruded rocks are characteristically in the form of pillow-lavas (described in Chapter III) and rest on a layer of peridotite (Figure 4). This is an ultrabasic rock (i.e. a heavy rock with a very high content of iron and magnesium) which commonly reaches the surface in a form known as serpentine. The lavas are overlain—especially towards the sides of the oceans—by thin deposits largely composed of the hard parts of minute floating organisms that fell to the ocean floor on death. From recent borings in most of the oceans of the world we know that these sediments are geologically young, that is to say, formed within the last 100 million years or so, and none is older than about 150 million years. This shows that the present oceans of the world were all formed since mid-Mesozoic times, though we know that parts of the continents are very much older.

Now this process of sea-floor spreading has been going on, not only in the Atlantic, but in all the oceans of the world (Figure 2).

The splitting of the crust normally, but not always, forms mid-oceanic ridges. In the eastern part of the North Pacific, for example, it runs into the coast of California and in the Middle East it forms the long narrow trough of the Red Sea and runs into the rift valley of east Africa.

Obviously we cannot have all the oceans of the world becoming wider and wider without the crust giving way somewhere. And give way it does, as we know from earthquakes. But the fascinating thing about earthquakes is that all the major ones are confined to certain parts of the world, and they all originate from planes sloping inwards and downwards from the edges of continents (Figure 4). So we see a mechanism for getting rid of the excess ocean floor—it is simply carried away down escalators, below the edges of the continents or island arcs. These 'escalators' are now known as **Benioff Zones** after the seismologist Hugo Benioff who first recognized them around the Pacific. The process of carrying oceanic material back into the interior of the earth is known as **subduction**. As the crustal material is carried down, deep trenches appear on the sea-floor (which commonly form huge sediment traps). So we arrive at a rather simple picture of convection systems operating mainly in the oceans and producing ocean floor material in the form of basic igneous rocks (i.e. lava flows etc., with a high iron and magnesium content). Just as hot water rises in a saucepan, or hot air rises in a debating chamber, so the lavas that rise rapidly at the mid-oceanic ridges show the mineralogical characters of high temperature and comparatively low pressure. Conversely, the ocean floor material being forced slowly down the Benioff zones is comparatively cool but is subjected to high pressure. Another complication known for some time in the oceans is that though the magnetic striping is regular, it is often displaced along lines parallel to the direction of movement (Figure 2). These displacements are known as **transform faults** and they play an important part in the jockeying for position of the continental plates.

So far we have only been considering plate tectonics in terms of continent-carrying plates being carried apart by a conveyor-belt mechanism of sea-floor spreading and subduction. But the other great benefit we have derived from these new ideas is what they tell us about the formation of the mountain ranges that figure so large in earth history. Plates can move apart, and they may slide past each other along transform faults, but they can also run together in the

1. Beinn Arkle, Sutherland from the south-west, showing the pale-coloured Cambrian quartzite above resting unconformably on the darker Lewisian gneiss which forms the low rounded hills in the foreground. Crown copyright

2. Intense crumpling in metamorphosed Dalradian limestone on sea-shore at Boyne Bay near Portsoy, Banffshire. From British Regional Geology handbook *Grampian Highlands*

3. Cader Idris near Dolgelly, Merionethshire. The cliff is over 1,000 feet high and consists of Ordovician volcanic rocks and sediments cut by a thick igneous intrusion (forming the obvious columnar face). The lake—Llyn Gader—lies in an old glacial cirque. Crown copyright

4. Ancient sun-cracks in the Old Red Sandstone near Thurso, Caithness; the cracks are filled with darker sediment which is more resistant to modern weathering and so forms the rims of pools. From *The Geology of Caithness*

5. A volcanic lava flow between two conglomerates in the Old Red Sandstone near Inverbervie, Kincardineshire. Crown copyright

6. A weathered surface of the Carboniferous Limestone near Penmaen, Glamorgan, showing segments of fossil crinoids. Crown copyright

7. Eglwyseg Mountain, the escarpment formed by the Carboniferous Limestone near Llangollen in Denbighshire. The rocks in the foreground and in the wood are of Upper Silurian age. Crown copyright

8. A normal fault in Coal Measures near Mossend, Lanarkshire. The massive sandstone bed at the top of the face on the left has been dropped down several feet on the right. Crown copyright

9. Zigzag folds in Upper Carboniferous sandstones and shales at Millock Haven in north Cornwall. In places the beds are truncated by short thrusts. From Bristol Regional Geology Handbook *South-West England*

10. Marsden Rock on the Durham coast, a modern sea-stack produced by erosion of well-bedded Magnesian Limestone in the Upper Permian. From British Regional Geology Handbook *Northern England*

11. The pebble beds in the Trias at Budleigh Salterton, Devon, overlain by weathered sandstone. Note the modern beach shingle in the foreground. Crown copyright

12. Hercynian granite on St. Mary's in the Isles of Scilly, showing the characteristic weathering into blocky 'tors'. The vertical and horizontal joints have no connection with the bedding seen in sedimentary rocks

13. 'False-bedding' of desert sand-dune type in Permian sandstone at Mauchline in Ayrshire. From British Regional Geology Handbook *The Midland Valley of Scotland*

14. Aust Cliff on the south bank of the River Severn near Bristol. The dark beds at the base and the prominent pale band belong to the Keuper Marl. These are overlain by the Rhaetian Series, and hidden in the vegetation at the top are the lowest beds of the Jurassic. Note the fault in the middle of the cliff which drops down the beds to the right, also the regularly cracked river muds in the foreground, which can be compared with the ancient cracks in Plate 4. From British Regional Geology Handbook *Bristol and Gloucester District*

15. The 'Seven Sisters' cliffs near Eastbourne in Sussex, showing the characteristic white Chalk of the Upper Cretaceous. Note the pebble beach in the foreground, entirely formed from flints derived from the Chalk. Crown copyright

16. Quarry near Erith in Kent showing the Upper Cretaceous Chalk overlain by the basal marine sands of the Palaeogene. Note the band of large flints at the junction. Crown copyright

17. Coast near Rhossili on the Gower peninsula, showing inclined beds of the Carboniferous Limestone planed off above by the supposed Pliocene marine invasion. The hill in the distance on the right is Rhossili Down formed of Old Red Sandstone in the core of an anticline. Crown copyright

18. Roche moutonnée of Dalradian schist plucked and smoothed by ice in Glen Nevis, Inverness-shire. From *The Geology of Ben Nevis and Glencoe and Surrounding Country*

19. Modern sand-dunes advancing across the countryside near Elgin on the Moray Firth. From British Regional Geology Handbook *Grampian Highlands*

20. Barchan sand-dune in the Sahara, southern Morocco. A characteristic horse-shoe shaped
dune developed in deserts. The prevailing wind comes from the right which has the gentler
erosional slope, whilst the points of the horse-shoe and the steeper depositional slope are
away from the wind. Note the characteristic stony desert surface (Photo, DVA)

21. Modern erosion on the Yorkshire coast near Aldbrough. The cliffs of Pleistocene boulder
clay are under active erosion by the sea, with disastrous results for the buildings and road
(Photo, DVA)

B

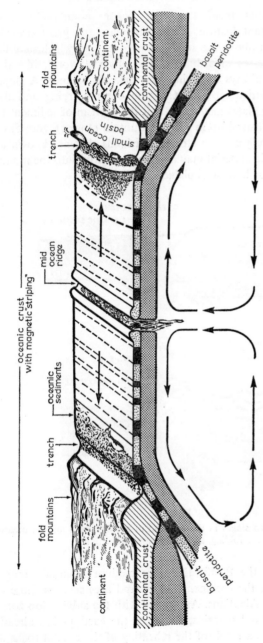

Fig. 4

Diagram showing the main elements of the plate tectonic hypothesis. Convection currents are shown producing splitting along the centre of an ocean. This leads to volcanicity and the formation of new oceanic crust and this is in turn carried sideways with 'normal' (black) or 'reversed' (stippled) magnetism. The ocean crust is consumed again down Benioff zones below a continent (left) or an island arc (right) with the formation of oceanic trenches. Oceanic sediments have accumulated most thickly on the older parts of the oceanic crust.

great collisions that form the long linear mountain ranges.

Another important feature of the earth are the great oceanic trenches that form above the subduction zones and which may fill with great thicknesses of sediment. The physical world, like the organic world, is always quivering in a delicate balance. A slight change in the relative movement of the ocean-carrying and the continent-carrying plates may cause that thick prism of sediment to be crumpled and heaved up in a new mountain range (Figure 5) or in one of those long curving strings of islands that are known as **island arcs**. Oceanic material may also be carried up into the mountains by the process known as **obduction** (Figure 6).

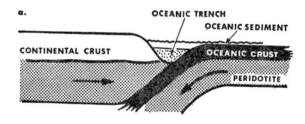

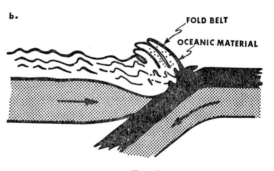

FIG. 5

Diagrams showing two stages in the collision of a continental and an oceanic plate producing orogenic folding, subduction and obduction.

Thus all round the Pacific there are modern mountain ranges along the edges of the continents and island arcs such as those of Indonesia and the Aleutians. Associated with the subduction zones below the continental margins, there are the earthquakes already mentioned and—as a result of the reheating of the crustal rocks as they plunge once more into the interior—there is volcanism and the emplacement of great masses of slowly solidifying molten rock. So we have what has been called the 'Pacific girdle of fire' all round that ocean. In places, such as Alaska and Mexico, the volcanoes are still very active, in others (such as the main part of the United States) the activity had largely ceased before the beginning of human history.

In the Atlantic we have a very different situation, with the two plates still separating along the Iceland line and each carrying both ocean crust and a continent. There are therefore no 'modern' mountain ranges, serious earthquakes or active volcanicity round the edges of this ocean and no marginal troughs either.

The third possible situation is seen in the Alps and especially in the Himalayas. These are comparatively 'modern' mountain ranges that have been formed by the crashing together of two plates carrying continents. So the Alps have been formed by the coming together of the European and African plates and the Himalayas have been formed by the collision of the Asian plate and that carrying the Indian subcontinent. This is discussed in Chapter X.

All this is what we see going on at the present time by means of magnetic, seismic and gravity surveys, from volcanism and earthquakes and from deep borings in the sea-floor. But more exciting still is the fact that we can extend the theory backwards into the geological history of the earth. Using the classic approach of uniformitarianism we can use these present-day processes to explain many of the mysteries that we find in the rocks. In other words, we can extend back the story that we can see being told today. Thus we can trace the North Atlantic back to when it was closed right up and when the sediments, the faunas and the floras in North America and Europe were very much alike, some two to three hundred million years ago, and these two continents were evidently very close together. But then we can go back even further, to a time when an earlier Atlantic was opening, roughly along the same lines. Thus in southwest Scotland we can find remnants of ocean crust that was sliding down a subduction zone nearly five hundred million years ago.

From the evidence of residual magnetism in the rocks we can see how earlier plates, often quite different from the present ones, moved about the surface of the earth. Oceans opened by sea-floor spreading and closed again with the consumption of oceanic crust down subduction zones. Deep troughs developed and filled with sediment above these zones and then were crumpled up against the neighbouring continental plates as new mountains. Slabs of plates slid past each other along transform faults and brought unlike structures alongside. Volcanoes burst out above the splits in the oceans and molten rock of a different kind squeezed up through the contorted rocks of plate collisions.

So we can see the process going right back in time into the Precambrian and geologists are now engaged in the exciting game of finding how the whole geological record fits in with plate theory. It must be said that so far, at least, it fits in very well.

II

THE OLDEST ROCKS OF ALL

*This was a difficult chapter to write since it attempts to summarize
the oldest and longest part of the geological record—the Precambrian
Era—which is represented by all those rocks which were formed
before the appearance of obvious fossils. It will probably also be a
difficult chapter to read, and some may prefer to tackle it later.*

If one sets out to write the history of ones village, one inevitably
finds that the oldest volumes of the parish register are difficult to
understand. They are written in an obscure script, they are
probably stained or damaged and vital pages may be torn or missing.
It is the same with the rock register which the geologist has to de-
cipher in order to write his history. The oldest part of the record is,
at the same time, the longest and the least well known. The Precambrian rocks are difficult to interpret because they have suffered
so much since their formation, they have been obscured by metamorphism, they have been damaged by later earth-movements and
(as with the parish register) many vital records have been lost.

What is more, in all the later rocks, there are fossil remains to help
us to correlate them from place to place. We know that the unfossiliferous Precambrian rocks are old merely because they look
old and because they are found beneath a pile of later records, but it
is as though the earlier vicars had not dated their registers and we
cannot be sure that some later absent-minded verger has not put
them back on the shelf in the wrong order.

The Precambrian is always associated in geologists' minds with
metamorphic rocks, because so much of the great Precambrian
areas of the world is occupied with rocks of this type. Indeed, for a
long time, any metamorphic rock was automatically classified as
Precambrian, and it was only realized comparatively recently that
rocks of any age may have been altered almost out of recognition by

37

heat and pressure. Conversely, unaltered sedimentary, volcanic and intrusive rocks are also found within the Precambrian succession.

SCOTLAND

The largest area of Precambrian rocks in Britain is in the highlands and islands of Scotland. The greater part of the mountainous area north of the industrial Midland Valley of Scotland is formed of these ancient rocks. The oldest rocks of all in the British Isles are known as the **Lewisian Series** after the island of Lewis in the Outer Hebrides which is composed almost entirely of rocks of this age. They consist mainly of **gneiss** (pronounced 'nice'), which is a coarsely banded metamorphic rock. Such gneiss is found in all the islands of the Outer Hebrides (see Figure 6) and in a narrow strip along the north-west coast of the Scottish mainland. Intense pressure and heat have made it very difficult to recognize the original nature of the rocks, but they seem to have been derived chiefly from older igneous rocks. There are some, however, which were certainly once sediment. These include the famous ornamental marbles of the islands of Iona and Tiree, west of Mull, which must originally have been deposited as sedimentary limestones. Another area of metamorphosed sedimentary rocks occurs around the head of Loch Maree on the mainland. Here the rocks are chiefly **schists**. This is another main category of metamorphic rocks, in which the constituent minerals are arranged in thin layers along which the rock readily splits. Those at Loch Maree include distinctive graphite schists full of the black carbon mineral graphite. In Finland the carbon in similar Precambrian rocks has been shown by its atomic constitution to be of organic origin. In much later rocks in the Alps, coal seams composed of plant remains can be seen to pass into graphite schists under metamorphic conditions. Therefore such occurrences may be evidence for some form of life (perhaps very simple plants) far back in Precambrian times. This possibility will be discussed later.

The main part of the Lewisian rocks probably originated as deep-seated igneous rocks of very varied composition. They are now seen as coarse-grained grey gneisses which form low rounded hills. In places there are concentrations of certain minerals which suggest an origin for that particular rock. Geologists used to speak of the

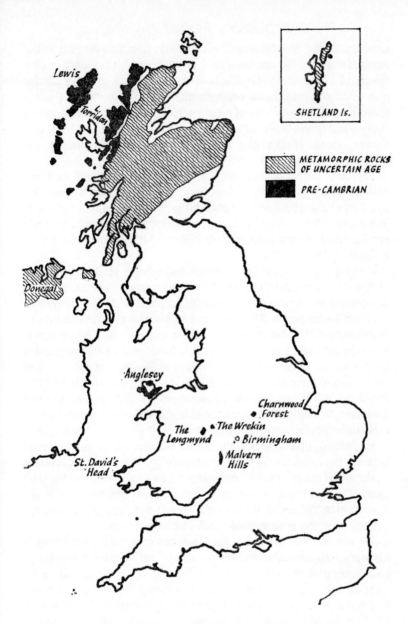

FIG. 6

Map showing the main outcrops of Precambrian and possible Precambrian rocks in Britain. Much of the latter are now known to be almost certainly Lower Palaeozoic in age.

Lewisian as the 'Fundamental Gneiss' and often thought of it as the original crust of the earth which first hardened on the surface of a sphere of liquid fire. We know now that the Lewisian rocks are far removed from the original crust. In other parts of the world there is succession after succession of metamorphic, sedimentary and igneous rocks which must be far older than any seen in the British Isles, and we are still far from getting down to the first crust, if indeed it is anywhere still preserved.

The Lewisian rocks of Scotland have been subjected to at least two periods of metamorphism; they have been broken and crumpled to a fantastic degree and have been cut by numerous dykes and other intrusions. They have also been very deeply buried and as deeply eroded again at least twice in their history. All this makes them very difficult rocks to study.

Resting on the Lewisian gneisses and schists in the north-west highlands are rocks of an entirely different type. They are unmetamorphosed sedimentary rocks—mainly thick sandstones—known as the **Torridonian Series**. The junction between this and the older rocks is interesting. The latter have a very irregular surface which represents the hills and valleys of an ancient landscape which was eroded by the winds, rains and rivers of many millions of years ago. The Torridonian is found in places filling old valleys as much as 2,000 feet deep. When one considers how deeply the Lewisian rocks must have been buried, and the amount of uplift and erosion necessary to raise these plutonic rocks and to remove all their overburden, then it is obvious that a very long period of time must have elapsed between the formation of the youngest rocks of the Lewisian and that of the oldest rocks of the Torridonian. The rocks themselves contrast markedly where they are seen in contact, and the bedding of the Torridonian (which is usually horizontal) cuts sharply across the structures in the metamorphic rocks (see Figure 9).

An obvious break of this kind between two rock formations is known as an **unconformity**. There will be many later examples of unconformities in our record, but hardly any of them will be so clear as this, nor will they represent anything like so long a break in the story. There are various types of unconformity dependent on whether the break is obvious or concealed and whether the bedding of the upper formation lies parallel to, or is at an angle with that of the underlying rocks. The surface between these two Precambrian formations is called a 'buried landscape'. The Lewisian rocks must have been exposed to the elements and worn away into the irregular surface we see now, before the Torridonian sediments were laid on top. There is evidence for this also in the nature of the sediments. They are of various types, but **sandstones** dominate the rest and form the high steep-sided mountains which characterize the Torridonian out-crop. As their name implies, these are rocks made up of sand-grains which are cemented together.

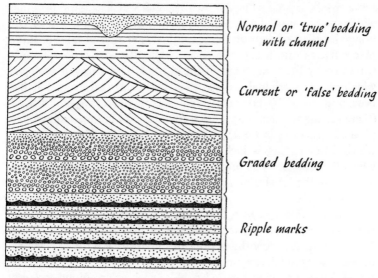

Normal or 'true' bedding with channel

Current or 'false' bedding

Graded bedding

Ripple marks

FIG. 7

Types of bedding in sedimentary rocks. Note the tendency of the 'false' bedding to be concave upwards, the graded bedding to become finer grained upwards and the ripple marks to be more angular upwards. These criteria are not com-pletely dependable, but false bedding particularly is very useful in determining the order in which beds were originally deposited.

The various divisions of the Torridonian change considerably in thickness from place to place; the maximum thickness for them all would be nearly 20,000 feet. This implies a very great deal of erosion of pre-existing rocks to provide 'such quantities of sand'. Various lines of evidence suggest that the sediment accumulated rapidly, probably under hot desert conditions. The sands are commonly red in colour, perhaps indicating an arid, oxidizing environment. They contain a great deal of felspar, a mineral which breaks down rapidly under normal erosion to form clays, so it is thought that the sediment was brought down by fast-flowing streams from a mountainous area not far away to the north-west. This is also suggested by what is known as **false-bedding** (or **current-bedding**—see Figure 7). 'Normal' or 'true' bedding in sedimentary rocks is parallel to the surface on which the sediments were laid down. Under special conditions, however, as in a delta, a shallow current-swept sea, a sand-dune or a mountain torrent, the sediment does not settle gently to the floor but is tipped over the front of an advancing heap. Thus the successive beds are inclined at an angle and we have 'false' bedding, which may give us the wrong impression of the inclination of the original surface of deposition. Often the lower ends of false-bedding planes flatten out in a gentle curve, whilst the tops are commonly truncated by the next sweep of the current. This tells the geologist which was the original bottom of a particular bed even if it has been overturned by later earth-movements. Such 'way-up' criteria (there are many others) have been found useful in unravelling complicated structures in the Highlands and elsewhere. The direction in which false-bedding is inclined may also indicate the direction in which the sediment was originally being transported. Such observations led to the conclusion that the Torridonian sediments came from the north-west.

There is other evidence that these sediments were laid down in shallow lakes or on dry land and not in the sea. At certain levels bedding surfaces show the imprint of raindrops and the cracks produced in wet mud drying under a hot sun. **Dreikanters** have also been found. These are wind-facetted pebbles such as are found today in deserts. They are produced by the wind blowing sand-grains against a pebble lying on the desert surface; the pebble occasionally rolls over and exposes a different surface to the sand-blasting action, so that it is eventually facetted on all sides.

Besides the sandstones in the Torridonian, there are some **shales** (well-bedded and compacted clay rocks), some **conglomerates** (cemented pebble-beds) and some **breccias** (which are like conglomerates except that the larger constituents are angular fragments and not rounded pebbles). The conglomerates were probably produced by torrential rushes of water carrying down boulders from the mountains. The breccias may be cemented scree material, suggesting that the mountains were close by. Long and careful searches have failed to reveal any fossils in these unmetamorphosed sediments, apart from some doubtful worm-tracks and plant spores.

In the Outer Hebrides there is one patch of supposed Torridonian around Stornoway in Lewis, which may in fact be much later in age. The mountain backbone of the islands is Lewisian and it may be that there were mountains in this area in Torridonian times providing the sediments which are found on the mainland.

To the south-east of the Torridonian mountains there is a major dislocation in the rocks known as the 'Moine Thrust'. This will be discussed later, but its importance in this chapter is as a boundary between the rocks just described and the very different ones on the other side. East of the Moine Thrust and north of the 'Great Glen', the greater part of the country is occupied by a complex set of metamorphic rocks known as the **Moine Series** or **Moinian**. Another great area of this series occurs beyond the 'Great Glen' in the Central Highlands, south of Inverness (see Figure 6).

The Moine Series has long been considered as Precambrian in age, but it is far from clear how it fits in with the Precambrian formations already described. On top of the Torridonian there are fossiliferous Cambrian sediments which will be considered in the next chapter. There seem to be three possibilities regarding the age of the Moinian:

1. It may be the equivalent of the Lewisian.
2. It may be the equivalent of the Torridonian.
3. It may belong between these two and have been cut right out of the record west of the Moine Thrust by the 'buried landscape' unconformity.

The Moine Series comprises a very varied succession of metamorphosed sediments which have been cut and impregnated by various igneous rocks. The former range from metamorphosed clay rocks—now seen as schists—to metamorphosed sandy rocks—now seen as **granulites**—which are particularly common south of the Great Glen. In places there are metamorphosed conglomerates and lime-rich rocks, and rarely, where the metamorphism is less intense, structures such as false-bedding can still be seen. In scattered areas in north central Sutherland there appear to be contemporary volcanic lavas and ashes which have been altered with the sediments.

The sediments of the Moine Series were penetrated by a series of igneous intrusions before they were metamorphosed. This is shown by the fact that the igneous rocks are themselves considerably altered. Most of them are **basic**, that is to say, they are dark rocks containing a high proportion of iron and magnesium minerals but no quartz. There are some extensive areas of **acid** rocks, at the opposite end of the chemical scale, which are pale in colour with a large percentage of quartz. These acid bodies are thought to have been emplaced at the same time as the metamorphism. They are not simple intrusions, but complex bodies produced by the injection of granite material into the schists.

The acid coarse-grained rock **granite** occupies large areas of the earth's surface and its origin has been the subject of considerable dispute. In some areas it seems to have the simple form of a fluid intrusion of molten magma which forced its way through the overlying rocks by simple mechanical means. In other areas, as in that described above, it seems to have permeated a pre-existing rock and to pass by inseparable gradations into ordinary gneiss. Some geologists regard granite as a magmatic igneous rock, others regard it as the product of a process known as **granitization**, in which an earlier rock is transformed by recrystallization and the expulsion of its more basic elements. Probably, as usual, both sides are parly right.

Within the Moine Series there are thought to be a number of **inliers** of Lewisian. An inlier is an outcrop of older rocks completely surrounded by younger, and may be produced in various ways. An **outlier** is the opposite of this—an area of younger rocks completely surrounded by older—and some of the isolated mountains of Torridonian sandstone perched on a foundation of Lewisian gneiss come into this category. Several of the supposed Lewisian inliers have in recent years been shown to be merely local variations within the Moinian, but others appear to be valid and imply that the Moinian is younger than the Lewisian.

This is just one of many arguments concerning the age of the Moine Series. Though dominantly sedimentary in origin, it differs from the sediments of both the Lewisian and the Torridonian. This may be explained by the Moinian sediments having been deposited in a different sort of area. Thus while the Torridonian sediments (which are very variable in thickness) were probably deposited under continental conditions at the foot of a mountain range, the Moinian sediments (which are much more uniform) appear to have been laid down in the sea, possibly in a marginal trough such as those discussed in the previous chapter.

The chief difficulties about the theory that the Moine Series is the equivalent of the Torridonian, are the metamorphic state of the former and the presence in it of many igneous intrusions. There is one area at the southern tip of Skye where the former difficulty may be partly solved. There are here rocks of Moinian type which are just west of the Moine Thrust and which are less metamorphosed than usual. This may indicate that the metamorphism decreased in intensity in this direction and that the metamorphic and non-metamorphic rocks were brought close together by later earth movements. Apart from this anomalous occurrence on the Sleat peninsula in the south of Skye, there are no Moinian rocks west of the Moine Thrust, and this seems to be a major argument against their being intermediate in age between the Lewisian and Torridonian.

In the southern part of the Highlands another major problem comes to join the confusion. This is the thick metamorphic series known as the **Dalradian** (see Figure 6). This occupies the greater part of the ground between the Moinian and the Midland Valley. Originally the Dalradian was a thick succession of sedimentary rocks, including conglomerates, sandstones, shales and **limestones** (i.e. rocks composed largely of calcium carbonate). These have been subjected to metamorphism of differing intensity, varying from low-grade metamorphism in the south to progressively higher grades in a generally northerly direction. A guide to the intensity of metamorphism is provided by certain minerals in the rocks which appear at progressively higher temperatures and pressures. Thus the micas are a group of minerals which appear at an early stage of metamorphism and can easily be recognized as glistening silvery flakes in the rocks, whilst tiny, reddish, almost spherical garnets appear at a slightly later stage.

Numerous local sedimentary successions have been worked out in the Dalradian Series and these have been correlated from place to place across large tracts of country, but the experts are still far from agreement on many major points. A great advance was made in our understanding of these rocks when it was found, as recently as 1930, that false-bedding and other criteria could be used to determine the order in which certain beds had been deposited. This led to the remarkable conclusion—ably demonstrated by Sir Edward Bailey—that over large areas of the south-west Highlands, the rocks have been completely inverted by later earth-movements. Apart from such major disturbances, the rocks have often suffered intense small-scale crumpling, as in the limestone shown on Plate 2. They were also disturbed by numerous igneous masses which appear to have been emplaced at about the same time as the metamorphism.

It is difficult to generalize about the setting under which the Dalradian sediments were first deposited, but their variety suggests a shallow-water environment where conditions were constantly changing. In places—for example in the fine sections exposed along the Banffshire coast—the sediments seem to have accumulated very rapidly. Certain of the 'boulder beds' in the Dalradian are particularly interesting. These are beds containing scattered pebbles which cannot be matched elsewhere in Scotland and are thought by some geologists to have been brought into the area by ice. There is not very much proof for this theory in Scotland, but the boulder beds have been compared with beds of undoubted glacial origin in the late Precambrian rocks of Norway and other parts of the world.

This brings one to the problem of the age of the Dalradian. Many geologists maintain that this series follows directly on top of the Moinian, and the two were long regarded as coming between the Lewisian and Torridonian in age, but this solution makes the section in the north-west Highlands even more difficult to understand. If the Moinian is the metamorphosed equivalent of the Torridonian, then the Dalradian may be a later division of the Precambrian not otherwise preserved in Britain. This fits in, to a certain extent, with the evidence from the thicker Precambrian successions in Norway.

Just before the last war, much interest was aroused by the discovery of Cambrian fossils near Callander in Perthshire, in a limestone which is regarded by some geologists as part of the Dalradian. It was therefore suggested that the Dalradian was all of early Palaeozoic age (see page 25) and not Precambrian at all. Much more recently, plant spores have been found which confirm this idea and the Dalradian is now generally accepted as representing, for the most part, metamorphosed early Palaeozoic sediments.

The study of the Scottish metamorphic rocks is almost a science in itself. Work continues there in a detail not attempted in any other part of the world, and the techniques and terminology used are scarcely intelligible to other geologists. But the complexity of structure and geological history that is being revealed shows (as detailed studies always show) that broad generalizations are not enough. The arguments continue, and will go on for a long time yet, but the result will undoubtedly be a knowledge of these rocks that far exceeds our knowledge of any other ancient part of the earth's crust.

THE OLDEST ROCKS OF ALL

IRELAND

The Dalradian rocks of Scotland seem to continue across the northern part of Ireland to the Donegal coast. There are other separated patches farther south. There are also several possible inliers of Lewisian rocks as in Scotland and they are all cut by some large granite bodies which emplaced at a later date.

ENGLAND

The Precambrian rocks of England are not to be compared in size of outcrop with those of Scotland and Ireland, but they include a very interesting variety of rocks. The most ancient are those of the Malvern Hills, which form a natural (though certainly not political) boundary between England and Wales. This narrow ridge of jagged hills has the Palaeozoic rocks of Wales on its western side and the Mesozoic rocks of England on its east. The greater part of the hills consist of gneisses and schists of great complexity but insufficient outcrop. They are known as the **Malvernian** and may be compared, though not reliably equated, with the Lewisian rocks of the Highlands. About half-way along the range is Herefordshire Beacon, below which there is an area of volcanic rocks which are younger than the metamorphics.

Similar volcanic rocks are much better seen in the nearest patch of Precambrian to London. This is in Charnwood Forest near Leicester, where a series of scattered pinnacles of volcanic lavas, ashes and other rocks project through a cover of much later sediments. There is no doubt as to their Precambrian age, because in a boring at Leicester they were found at depth, overlain by fossiliferous Cambrian sediments. A few years ago a definite fossil, looking like a leaf, was found in these rocks. It is probably a very simple marine animal and is very similar to specimens in quite a rich late Precambrian fauna known from South Australia.

The finest area in England for studying the later Precambrian rocks (and much else besides) is in the geologists' playground of Shropshire. Here Precambrian volcanic rocks are well seen in the

Wrekin, which overlooks the little town of Wellington. Nearby is the buried predecessor of Shrewsbury where A. E. Housman's 'Roman and his troubles, are ashes under Uricon'. The vastly older ashes and lavas of the Wrekin are known as the **Uriconian.**

A long line of similar narrow, hog-backed hills extends down past the old market town of Church Stretton, which stands under the largest of them—Caer Caradoc. Other patches of Uriconian volcanic rocks come to the surface farther west, notably in Pontesford Hill near Minsterley. It is interesting that a thick volcanic series should be developed in the Precambrian of southern Britain, but not in the classic sections of the western Highlands. It should be remembered, however, that there are volcanic rocks in the Moinian and some are also known in the Dalradian. This may have a bearing on the relative ages of the various series.

On top of the Uriconian in Shropshire comes a very thick sedimentary succession of conglomerates, sandstones and shales. This is the **Longmyndian Series**, named after and forming the great upland region of the Longmynd, which rises west of Church Stretton (see Figure 16). The rocks here are steeply inclined and show puzzling relationships to one another and to the other rocks of the neighbourhood. There has been much argument about them, but it seems certain that as a whole they are younger than the Uriconian but older than the Cambrian. They have often been compared with the Torridonian of Scotland, but there is no real validity in attempted correlations of this sort. A similar series is exposed in a small area near Ingleton in west Yorkshire.

The only other possible area of exposed Precambrian rocks in England, is in the Cornish peninsula of the Lizard. Here there is a confused set of mainly basic metamorphic rocks, which may be, and probably are, much later in age.

WALES

The great place in Wales for the Precambrian is the extreme north, especially on the flat-topped island of Anglesey which is largely composed of such ancient rocks. A very thick series of varied sediments, somewhat like the Dalradian, rests unconformably on a very complex series of highly metamorphosed rocks comparable to the Lewisian.

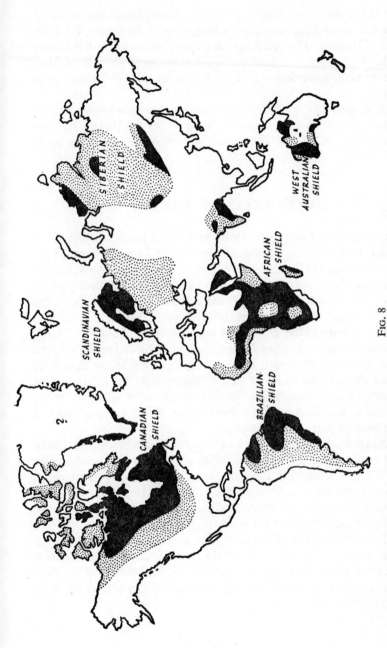

FIG. 8

Simplified map showing the shield areas of the world. Large regions in which Precambrian metamorphic rocks are exposed at the surface are shown in black; platform areas, in which these rocks are buried at shallow depths beneath unfolded sediments, are shown stippled.

THE OLDEST ROCKS OF ALL

On the mainland opposite Anglesey, the Precambrian comes up again in two ridges, one between Bangor and Caernarvon, the other near Llanberis. The rocks here are partly sedimentary but mainly volcanic and are reminiscent of the Uriconian of Shropshire. They are overlain unconformably by fossiliferous early Palaeozoic sediments.

Volcanic rocks of Uriconian type are again seen below the Palaeozoic in the south-west corner of Wales—near St. David's in Pembrokeshire. In all the Welsh areas the layered Precambrian rocks are cut by many igneous intrusions.

OTHER PARTS OF THE WORLD

Precambrian rocks cover great areas of the earth's surface. They occur particularly in huge **shield areas** which form the cores of the various continents (see Figure 8). In Europe we have the Scandinavian Shield, centred on Finland, around which the continent grew up like layers of straw and earth round a clump of potatoes. A similar shield in Siberia forms the nucleus of Asia (with a smaller shield in southern India). The greatest of them all is the Canadian Shield, which outcrops over some 2,000,000 square miles and contains a vast amount of mineral wealth, as does the African Shield, which extends from the extreme south of that continent right up to Arabia. In South America there is the Brazilian Shield, split into two by the Amazon, and in Australia a Precambrian shield occupies most of the western part of the continent.

Apart from the shield areas of exposed Precambrian rocks, there are surrounding platform areas—such as that under the western part of Russia—where the Precambrian is buried at no great depth beneath flat-lying sediments. There are also smaller areas, as in Britain, in which the Precambrian appears where the rocks have been strongly folded in the hearts of old mountain ranges.

The vast majority of the world's mineral wealth in metal-bearing ores lies in the ancient rocks of the shield areas, and these receive the chief attention of mining geologists. On the other hand the coal-bearing rocks are much later in age and the majority of the oil-bearing formations are later still.

DATING THE ROCKS

In the mid-seventeenth century Dr. John Lightfoot, vice-chancellor of Cambridge University, announced that—as a result of careful study of the Bible—he had determined that the earth was formed '. . . on the 26th October 4004 B.C. at 9 o'clock in the morning'. Such authoritative statements were generally accepted, and two centuries later J. W. Burgon could still call Petra, 'A rose-red city half as old as time' and really mean it.

Right up until the later part of the last century, intelligent men believed quite sincerely that the world was less than 6,000 years old, but as knowledge accumulated, geologists demanded more and more time for the accomplishment of all the events they found recorded in the rocks. By the eighteen-sixties they were already requiring hundreds of millions of years for all that had happened since the formation of the earth.

Today the estimates increase at a fantastic rate from year to year. The largest estimate of age known to the author at the time of writing is for some rocks from west Greenland which are thought to be 3,725,000,000 years old (with a possible error of 25 million years in either direction). This estimate was for some igneous rocks which had been intruded into sediment. Thus we have to go back two more stages, through the period of deposition of the sediments and the period of formation of the rocks which were eroded to form the sediments. Even then it is probable that yet older rocks will be discovered and we shall still not know how far we are from the original crust of the earth.

The uninitiated always demand to know how scientists can possibly estimate such unimaginable lengths of time, and they are rarely satisfied with the answer. This is a matter for the chemist rather than the geologist, and concerns the gradual decay of those notorious but inestimably valuable substances, the radio-active elements. For many years elements were regarded as immutable, but the modern alchemy of nuclear physics has shown that certain elements, such as uranium, break down at a fixed rate to form stable elements such as lead. Some do it very, very quickly, but a number are now known which do it slowly enough to be useful in dating rocks. The rate of decay of radio-active elements does not seem to be affected by any possible

51

range of heat or pressure, and it is a reasonable assumption to suppose that it has gone on at the same rate throughout earth history. Therefore, if a uranium mineral is found, part of which has obviously changed to lead, it is possible to calculate from the amounts of the two minerals present, how long it is since the older mineral was first emplaced in the rock. Various errors can creep into the calculations, but when the evidence is clear (which is not often) the amount of possible error can be calculated with reasonable accuracy. Elements such as uranium and thorium are useful in dating the older rocks. Recently a great deal of work has been done on dating events in the last few thousand years by means of a radio-active variety of carbon. This will be discussed in Chapter XI.

The pious Victorians found it difficult to accept the geologists' concept of earth history, just as the Catholic Church centuries before had rejected Galileo's concept of the universe. In both cases, the fundamental cause of the discomfort was the same; the new ideas reduced man and all his history to an extremely insignificant place in the physical world. Today we accept this position from our earliest intelligent years and can hardly understand what a shock it was for our grandfathers.

III

THE OLDEST ROCKS WITH FOSSILS

This chapter is concerned with the three periods of geological time known as Cambrian, Ordovician and Silurian (see table on page 25). These three periods form the first half of the Palaeozoic (or Primary) Era and are commonly spoken of together as the Lower Palaeozoic.

THE CAMBRIAN PERIOD

All along the north-west coast of Scotland, from Skye to Cape Wrath, the cliffs are formed of the old Precambrian rocks described in the last chapter—the red Torridonian sandstones and the ancient Lewisian gneiss. But if one goes ten to twenty miles inland, up one of the numerous locks or across the bare moorlands, one sees on the mountain tops rocks of a different type, which mark the beginning of the next episode in our history.

Just south of Loch Assynt in Sutherland is the mountain of Canisp. In this mountain is clearly recorded the whole early history of Scotland (see Figure 9). The mountain rises from a rolling plateau of grey Lewisian gneiss, which projects in innumerable little crags from the thin turf. Higher up are steep-sided cliffs of the dark red Torridonian sandstone which forms most of the mountain. At the top of the western face are white, sandy sediments which are sometimes mistaken for snow. These were laid down, not on the land like the Torridonian, but in a shallow sea which invaded this area some six hundred million years ago. This invasion was the first of several marine transgressions which we shall meet in the course of our history. Before this, there must have been a long period during which the older rocks were being worn away by wind, rain and running water. Then the sea slowly encroached upon the land, eating away the Precambrian rocks in the same way as a modern sea wears away

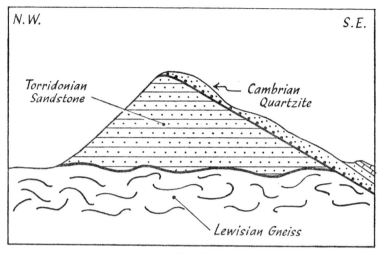

FIG. 9

Simplified cross-section of Canisp in Sutherland, showing the Cambrian beds (on the right) resting on Torridonian and Lewisian rocks of the Precambrian.

its cliffs. As the cliffs moved back, they left a flat surface under the sea on which the first Cambrian sediments were deposited. So today the Cambrian strata are seen to rest on a planed-off surface of older rocks. In places the whole of the Torridonian has been removed below this surface and the Cambrian rests directly on the Lewisian gneiss (as on the right of Figure 9 and in Plate 1). This is an excellent example of an 'unconformity', like that between the Torridonian and the Lewisian described in the last chapter. Both are seen on Canisp, but whereas the lower unconformity is a 'buried landscape' with the Torridonian sandstones blanketing the old Lewisian mountains and valleys, the upper one is a 'plain of marine erosion' with the marine Cambrian sediments cutting across everything.

Another special feature of the base of the Cambrian is the presence of a thick bed of rounded pebbles. This represents the first shingle beaches of the Cambrian sea as it spread across the old land-mass. Such rocks, made up chiefly of pebbles, are conglomerates, and are often found marking marine transgressions. Above the conglomerate comes a thick series of sediments known as **quartzites**; these are hard sandstones in which the sand-grains are cemented together by quartz, so that the rock is almost pure silica. Such quartzites are found almost everywhere in Britain at the base of the Cambrian.

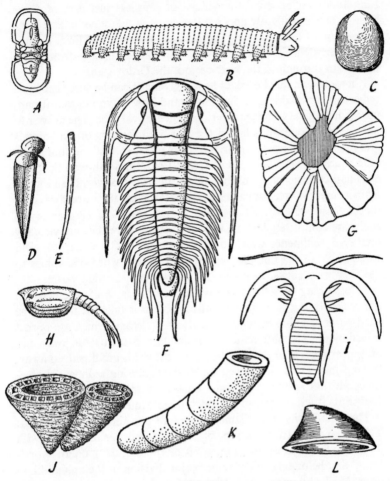

FIG. 10

Cambrian fossils. *A*: small blind trilobite, × 3½; *B*: reconstruction of a worm-like creature from the Burgess Shale (after Walcott), × 1⅔; *C*: simple horny brachiopod, × ¾; *D* and *E*: members of an extinct group, possibly related to the molluscs, both × 1; *F*: large spiny trilobite, × ½; *G*: jelly-fish from the Burgess Shale (after Walcott), × ⅔; *H*: crustacean distantly related to the lob-sters, × ¾; *I*: aberrant crustacean from the Burgess Shale (after Walcott), × 2; *J*: extinct group with points in common with both the sponges and the corals, × ⅔; *K*: early cephalopod mollusc, × 1½; *L*: early marine gastropod mollusc or 'snail', × 2. The magnifications given for these and later drawings of fossils are only approximate.

THE OLDEST ROCKS WITH FOSSILS

They are followed in the north-west Highlands by the 'Durness Limestone', called after the village of Durness just east of Cape Wrath. This is a slightly unusual limestone in that it contains a large proportion of magnesia besides the usual carbonate of lime. It must have been deposited very slowly, and ranges in age from early Cambrian through to the next period, the Ordovician.

It may reasonably be asked how a geologist can be sure that these sediments were laid down in the sea and how he knows that it happened during the Cambrian Period of geological time. The answer to both of these questions lies in the fossilized remains of sea animals and plants which are found in the rocks just described, but not in the older ones below. The types of fossils found in north-west Scotland are characteristic of Cambrian rocks all over the world, and apart from a very few exceptions (mostly controversial) these are always the oldest fossils known. Almost everywhere the older (Precambrian) rocks are completely lacking in fossils, even where they are virtually unaltered sediments which one might reasonably expect to yield some traces of organic life. On the other hand, when one comes to the Cambrian rocks, fossils are found in quite startling variety.

A fossil may be defined as any record of organic life preserved in the rocks. In the vast majority of cases, fossils take the form of the hard parts of animals, such as shells or skeletons, which are buried in sediment. The soft tissues of animals and plants rot away and disappear soon after death. Often even the shells are dissolved away, leaving just a cast or impression in the enclosing sediment. Sometimes all that is known of some long-dead animal are the trails or footprints it left on crossing a patch of soft mud.

The chief fossils of the Cambrian are of three kinds (see Figure 10): **Trilobites:** an extinct group of many-legged sea creatures with jointed external skeletons, somewhat resembling the modern 'wood-louse' and belonging to the same major division of the animal kingdom; **Molluscs:** comparable to those found in our modern seas, but only poorly represented, chiefly by univalved forms (**gastropods**, like the limpet and the whelk); **Brachiopods:** a group of bivalved sea creatures, still living in a humble way at the present day, but much more important in the past, and not to be confused with the bivalved molluscs.

These groups of fossils are represented by scores of genera and species in the Cambrian rocks, though they are known, almost exclusively, only by their hard parts. By careful study of these remains,

56

experts can see that there was not just one Cambrian fauna, but several, which succeeded each other in time. Thus we can subdivide the Cambrian rocks into Lower, Middle and Upper Cambrian on the basis of their fossils, and within these large divisions it is possible to recognize many smaller subdivisions or zones. In this way the strata of one age can be correlated from place to place with fair accuracy. Thus we can say, with some confidence, that a certain set of beds in Wales was laid down at about the same time as certain beds near Oslo in Norway. The margin of error is considerable, perhaps millions of years in the older strata, but the accuracy increases in later beds where there are more abundant and more suitable fossils. It depends very much on the lateral distance between the rocks to be correlated and the nature of the fossils available.

An important point to remember about fossils is that they always give us a very distorted picture of the life present on earth at any particular time. Only a tiny proportion of the organic world has the right structure and lives in the right situation to have a chance of being preserved; even then the likelihood of any particular animal or plant being fossiilized is extremely remote. So it is that the larger, more surprising and more publicized fossils such as the 'dinosaurs' of later times are very rare, whilst humbler fossils—'sea-shells' and the like—are much more common and are in fact the everyday tools of the geologists.

Nevertheless, scattered up and down the stratigraphical column in various parts of the world, are little local miracles of preservation where the most unlikely organic remains have survived in the rocks. One such 'miracle' was found in the Middle Cambrian rocks of British Columbia in western Canada early in the present century. The deposit—known as the 'Burgess Shales'—occurs at the top of Mount Stephen where the Canadian Pacific railway cuts through the Rockies. The fauna was found and studied by the great American palaeontologist C. D. Walcott, who filed a mining claim on the property and literally quarried the rock at the top of the mountain, lowered it on sleds and then carried it off on pack-horses to his laboratory to be carefully split up and examined. The Burgess Shales must have been laid down on the bottom of a very quiet sea in which the remains of the multitudes of sea creatures were not broken up by waves and currents. The muds on the sea-floor were stagnant and too unhealthy even for the ubiquitous bacteria which are essential for the processes of decay. As a result, the dead bodies of animals which sank down into the mud did not rot away, but are still to be found perfectly preserved in the very fine-grained shales which the muds have become. Though only known from this one locality, the fossils of the Burgess Shales give us a remarkable picture of the variety of life already in existence in Cambrian times. Besides the animals with hard skeletons such as the trilobites, there were many soft-bodied forms such as worms, sponges, jelly-fish and a great variety of crustaceans; in fact there were representatives of most of the major groups of invertebrates living today and several extinct groups besides. This great variety of Cambrian sea life leads one to wonder how it can have developed so soon after the first appearance of life in the rocks.

The origin of life is perhaps the most difficult problem which the scientist still has to solve—if indeed it lies within his field at all—but it is a problem for the biologist, the organic chemist and the theologian rather than for the geologist. All that the geological record can tell us is that definite signs of life first appear in the rocks which we call Cambrian. In the Precambrian rocks there is, in a few places, some evidence of very simple plants and some very rare and often disputed animal fossils; there is also some indirect evidence of life from deposits of carbon, lime and ironstone which seem to imply organic action. Furthermore, unless we accept a single act of creation, a complex organism such as a trilobite must have had a long previous history of gradual evolution, and this is even more necessary for the varied fauna of highly specialized animals seen in the Burgess Shales.

However, all this evidence is very slender and indirect, and the chief problem for geologists is why fossils suddenly appear, in comparative abundance, at the same stratigraphical level all over the world. Many explanations have been suggested, some reasonable, some fanciful. Early life may have evolved in an isolated marine basin, not so far discovered, and from which it only escaped to colonize the rest of the world at the beginning of Cambrian times. Alternatively it may be that this was the time when animals first developed the ability to secrete a hard shell, either because the earlier seas lacked the necessary chemicals or because there was suddenly the need for protection from a newly-evolved predatory enemy. Another idea is that life originated in the oceanic depths, and any deposits which were laid down there have remained under the sea (and so unexplored) ever since. The beginning of the Cambrian, by

Durness

L. Assynt

CAMBRIAN

The Wrekin

Nuneaton

Harlech

Lickey
Hills

St. David's
Head

FIG. 11

Map showing the main outcrops of Cambrian rocks in Britain. The Cambrian age of the outcrops shown in south-east Ireland and the Isle of Man has not yet been proved.

this theory, was then the time when life first colonized the shallow waters which are represented in the sediments we can now study on dry land. Whatever their origin, from the Cambrian onwards, fossils are the calendar by which our history is dated.

In Britain, besides the occurrences in north-west Scotland which have already been described, Cambrian rocks are found in Wales and in the English Midlands (see Figure 11). The name Cambrian comes from the Roman name for Wales, where rocks of this age outcrop over large areas and where they were first studied and named by the Cambridge pioneer Adam Sedgwick. They are especially well displayed around Harlech, at the north end of Cardigan Bay. Here there is a large, oval area of Cambrian rocks (see Figure 12) known as the Harlech Dome, for the rocks are folded in the form of a huge dome, sloping outwards in all directions. As a result of long ages of erosion, the centre of the dome has been deeply pierced, revealing the oldest rocks in its core. The rocks consist chiefly of great thicknesses of grits (coarse sandstones with sharp, angular grains) and slates. The grits are harder and so have weathered out as ragged, barren mountains, whilst the slates now form low-lying country in between.

If one takes the lonely road northwards from Dolgelly towards Blaenau Ffestiniog, one comes to a flat area in the centre of the dome, like a vast amphitheatre. On the low ground can be seen exposures of the lowest Cambrian slates of the area, whilst all around is a ring of mountains formed by the lowest grit formation. The top part of the Cambrian succession is mainly of slates and is overshadowed by the surrounding mountains of Ordovician volcanic rocks—Snowdonia to the north, the Rhobell Fawr mountains to the east and the Cader Idris range to the south. Another smaller area of Cambrian rocks is found north of Snowdonia, around Llanberis, where they are mainly slates. From the great Penrhyn quarries come the famous purple and green lower Cambrian slates which were formerly used for roofing all over the world. Cambrian rocks are also found in South Wales—in north Pembrokeshire—where giant Middle Cambrian trilobites, up to three feet long, can be found on the coast near St. David's Head (if one knows the right place to look).

The English Cambrian is only a shadow of the great thicknesses of sediment seen in Wales. In Shropshire it is found in a series of tiny scattered patches. The most spectacular development is that in the Wrekin range, just south of Wellington. These hills are formed

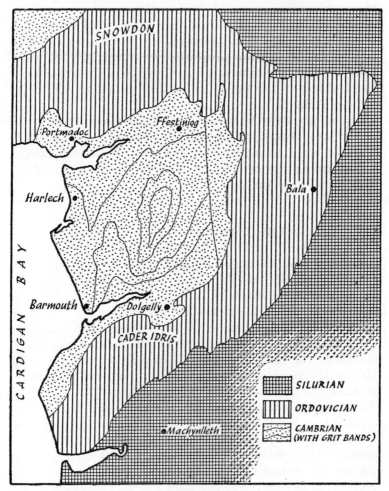

FIG. 12

Geological sketch-map of part of North Wales, showing the Harlech Dome and surrounding areas (simplified from various sources).

mainly of Precambrian volcanic rocks, but if one looks back as one climbs the path to the highest point, one sees the gleaming white basal Cambrian quartzite plastering the east side of the ridge. In the large quarry in Buckatree Glen, this quartzite can be seen resting unconformably on the Precambrian with a conglomerate at its base —very much as in north-west Scotland. Above the quartzite in

61

Shropshire come conspicuously green sandstones—green because of the presence of an iron mineral called glauconite which is a sure sign of marine deposition. There are also shales, which are muddy sediments that have not been compressed and altered as much as the slates of Wales. The same rocks are found again farther south—in Herefordshire—where they lie against the Precambrian ridge of the Malvern Hills.

Farther east, the white Cambrian quartzite forms the greater part of the Lickey Hills, a popular place of recreation near Birmingham. It is seen again, resting on the Precambrian, in the long series of huge quarries near Nuneaton, where it is broken up to be used as a foundation for railway tracks.

THE ORDOVICIAN PERIOD

The most obvious change which occurred at the beginning of the Ordovician Period in the British area was the sudden bursting into action of many great volcanoes. All round the Harlech Dome in North Wales are volcanic mountains of Ordovician age—not actual volcanoes, for these became extinct and were eroded away long ago —but great heaps of volcanic material interspersed with ordinary sediments. The north face of the Cader Idris range, south of the Harlech Dome, has at least eight distinct beds of volcanic rock— ashes, agglomerates (volcanic pebble-beds) and lava flows (see Plate 3). At the summit is a lava flow of a special sort, usually called a pillow lava. It is thought to have been extruded under the sea. As the hot lava welled up from a fracture in the sea-floor, it cooled in contact with the water and formed a hard crust. The pressure continued until the lava burst through this crust and the process was repeated. In this way there was formed a series of pillow-like masses of lava, each with a crust which shows signs of rapid cooling. Similar lava-flows can be seen forming off the Hawaiian Islands at the present day. They are particularly associated with the ultrabasic rocks known as serpentines and with deep-water deposits full of the remains of minute planktonic organisms. All these are indicative of oceanic rather than continental crust, and the distribution of Ordovician pillow lavas, etc. in Britain has led to the suggestion that we have here the remnants of at least two subduction zones (see Figure 15) on the opposite sides of what may then have been a very wide ocean (sometimes called the Proto-Atlantic). Most of the floor of that ocean was later consumed down these subduction zones, so bringing the two sides close together.

Volcanic activity went on intermittently all through the Ordovician. Volcanic rocks are found between the ordinary sediments all down the Ordovician outcrop into South Wales where, for instance, they form the greater part of the bird sanctuary island of Skomer (see Figure 13). Associated with the extrusions of lava and ash which poured out at the surface, were intrusions beneath the surface. In the latter, molten rock was forced into the older sediments, either as vertical sheets (dykes) along fractures or as horizontal sheets (sills) along the bedding (see Figure 1). Also associated with these intrusions are the much-sought mineral veins, carrying metal ores of economic importance. Ores such as those of lead and zinc were formerly mined in many Ordovician volcanic areas, but few of them are rich enough to be worth mining today when they can be worked on a much bigger scale in other parts of the world. Even gold is to be found, and North Wales is the traditional source of gold for royal wedding-rings.

Ore deposits of this type were formed later than the rocks in which they are found. They are often seen filling cracks and so are very thin, though they may go down to great depths. In other cases they may impregnate the parent rock, often filling all the interstices, for example between the grains of a sandstone. Other mineral deposits are found as irregular bodies, which completely replace the original rock and grade off into it imperceptibly in all directions. All these deposits clearly come from below and appear to have originated from gases or liquids rising from a molten magma just before it solidified. The veins are not often composed entirely of the valuable ore. Usually they are formed mainly of some less valuable mineral such as quartz, and veins composed of quartz alone are very common in all areas of ancient rocks. They can be seen as thin white bands cutting across the strata and occasionally opening out around cavities into which the quartz projects as almost perfect crystals.

As one goes eastwards into England, the evidence of Ordovician volcanicity diminishes. In west Shropshire—around the village of Shelve on the Welsh border—there are still several volcanic levels in the Ordovician, though on a much smaller scale than in Snowdonia and Cader Idris. There are also big intrusions and the ruined mine buildings of many old workings. In east Shropshire, however, near

SILURIAN

ORDOVICIAN

FIG. 13
Map showing main outcrops of Ordovician and Silurian rocks in Britain.

the little town of Church Stretton, evidence of volcanicity has disappeared completely from the Ordovician rocks and there are no mineral veins. Farther east still, there are no Ordovician rocks known at all (apart from one borehole in Cambridgeshire) and it is thought that the seas of this age never spread over the Midlands. But the Cambrian rocks here are cut by intrusions which do not go up into the Silurian above and these are blamed on the Ordovician volcanic disturbances.

The Lake District, in north-west England, is also largely built of Ordovician rocks, and a large proportion of these are volcanic or intrusive. German miners came here during the reign of Elizabeth I to look for useful minerals and a prosperous mining industry flourished for many years. Lead, copper and graphite (for pencils) are among the minerals that have been worked here in the past.

In Scotland too, there was intense volcanicity during the Ordovician Period and we have near the seaside town of Girvan in the south-west, the obvious remnants of the subduction zone that carried the crust of that early Atlantic down under what are now the Scottish Highlands. In the north-west Highlands, the top of the Durness Limestone (see page 56) contains an early Ordovician molluscan fauna exactly like one found in North America, though unlike those found elsewhere in Britain. This and the Cambrian beneath show a marked contrast, both in rocks and fossils, between those of Scotland and those of England and Wales. This is clear evidence of the wide ocean that separated them in early Palaeozoic times. Above the Durness Limestone the record is broken and nothing is known of the rest of the Ordovician or the Silurian north of the Midland Valley of Scotland. To the south of that valley, however, are the Southern Uplands of Scotland, which are composed largely of Ordovician and Silurian sediments. These are thought to have accumulated in a marginal trough above the subduction zone.

In 1866 a schoolmaster, Charles Lapworth, started studying the Lower Palaeozoic rocks around Moffat in the centre of the Southern Uplands. He was a determined worker and chose this area because it was one of great geological complexity, which had not then been unravelled. He spent more than ten years tramping the peat-covered hills and steep-sided valleys before he published his results and established himself as one of Britain's greatest geologists. The chief tool he used in his researches was a group of then little-studied fossils known as **graptolites** (see Figure 14). These were small, plant-like,

C 65

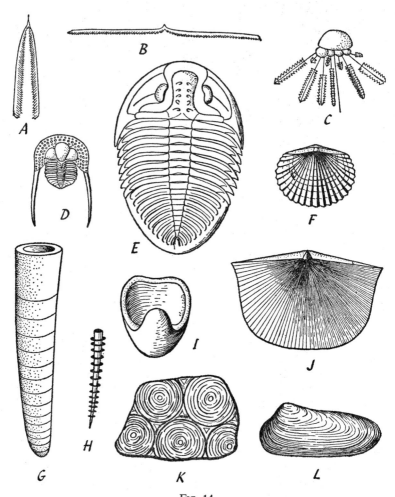

FIG. 14

Ordovician fossils. *A, B* and *C*: different kinds of graptolites, *A* × 1, *B* × ⅔, *C* showing several graptolite colonies attached to a float (after Ruedemann) × ½; *D* and *E*: trilobites, showing the characteristic trilobed appearance of the skeleton, *D* × ¾, *E* × ⅔; *F* and *J*: brachiopods with calcareous shells, both × 1; *G*: nautiloid cephalopod with straight shell, × 1; *H*: member of an extinct group, possibly related to the molluscs, × 1; *I*: marine snail or gastropod, × 1½; *K*: rock formed by lime-secreting algae, × ½; *L*: bivalved mollusc or lamellibranch, × 1½.

colonial animals which were so called because of their resemblance to pencil-markings on the black slates. Graptolites are very abundant in Ordovician and Silurian rocks and are ideal time-indicators, for they were widely distributed and evolved rapidly. They probably floated near the surface of the sea attached to seaweed or to special floats; their dead remains accumulated in the poorly-oxygenated muds on the sea-floor which could not support normal sea-floor life. In the great gloomy valley of Dobb's Linn, east of Moffat, Lapworth found different kinds of graptolites every few inches in the dark slates. The succession of forms that he observed along one stream he found repeated on all sides, and so by means of the graptolites he was able to work out the very complicated folding and faulting which had affected the rocks of this neighbourhood.

Later Lapworth extended his studies north-westwards and made another classic investigation, this time of the area around the small town of Girvan on the Ayrshire coast. Here he again found Ordovician and Silurian sediments, but of a very different nature from those at Moffat. At Moffat, the Lower Palaeozoic rocks consist of slates and grey muddy sandstones. At Girvan there are limestones, conglomerates and other sediments suggestive of the sea-shore. What is more, the majority of the fossils are also different. Whereas at Moffat graptolites are super-abundant to the virtual exclusion of everything else, at Girvan there are brachiopods, trilobites, molluscs, starfish and many other forms, besides enough graptolites to enable a correlation to be made between the two areas. It appears that at Girvan we are close to what was the shoreline of those days, whereas at Moffat we are far from that shoreline, apparently in the middle of a great rapidly-sinking trough (i.e. an oceanic trench).

A similar distribution of sediments and fossils can be seen farther south. In Wales, the main outpourings of volcanic material apparently occurred near the centre of a trough, for they are associated with slates full of graptolites. In the Welsh border counties of England (see Figure 15) there are limestones, pure sandstones, limy shales and an abundance of varied fossils as at Girvan. This then would seem to be the shoreline, which is confirmed by the disappearance of Ordovician sediments farther east. Near Builth, on the Welsh side of the border, geologists discovered an actual shoreline of the Ordovician sea, still preserved, with shingle-beaches, inlets and sea-stacks.

So we can trace one oceanic trough (or **geosyncline**) running roughly south-west to north-east up through Wales and the Lake District. Another geosyncline, or a continuation of the same one, extended from north-east Ireland through the Southern Uplands of Scotland, perhaps reaching as far as Scandinavia. There are many such geosynclines up and down the stratigraphical column in various parts of the world. The name is generally applied to long, comparatively narrow belts in which sedimentation (usually of a distinctive kind) is very thick, but thins markedly towards the margins. On one side of them there were often wide shallow seas, on the other side we now think that there were only the thin deposits of the oceanic depths. This does not necessarily imply very deep sea troughs at any particular time, but merely that the earth's crust along these belts was subsiding rapidly and constantly, and so receiving an undue amount of sediment. The 'Anglo-Welsh geosyncline', as it is called, which is thought to have persisted in that region all through Lower Palaeozoic times, must have accumulated a thickness of something like seven miles of sediment, compared with only a few hundred feet along its margins. One particularly important feature of geosynclines is that they always determine the line of later mountain ranges, but this matter will be discussed later.

It will be seen from the above that by studying all the available geological evidence, it is possible to deduce, or at least to guess at, the geography of a particular region at a particular time in the past. Much energy goes into the production of such **palaeogeographical maps**, and several simplified ones are given in this book (see Figure 15 for the Ordovician map). The degree of possible error, however, is considerable, especially for the earlier periods, and they should never be taken too seriously.

The Ordovician was the last of our geological systems to be named. Adam Sedgwick, the Cambridge professor who named the Cambrian System, worked upwards from below. Roderick Murchison of the Geological Survey in London, named the Silurian System and worked downwards. Inevitably there was an overlap in the strata studied by these two great pioneers, and there ensued one of those bitter controversies over nomenclature which afflict all sciences at one time or another. It was not settled until both contestants were dead and Charles Lapworth proposed the Ordovician System for the disputed strata. He named it after the Ordovices—the last and fiercest of the British tribes who fought against the Roman invaders and who lived in an area where these rocks are particularly well developed.

THE OLDEST ROCKS WITH FOSSILS

It should not be thought, however, that this system was merely a compromise, for it records, both in its fossils and in its geological history, a very distinct and important period of time.

THE SILURIAN PERIOD

The Silurian Period followed the Ordovician in the British area without many marked changes. Volcanic activity soon died away to

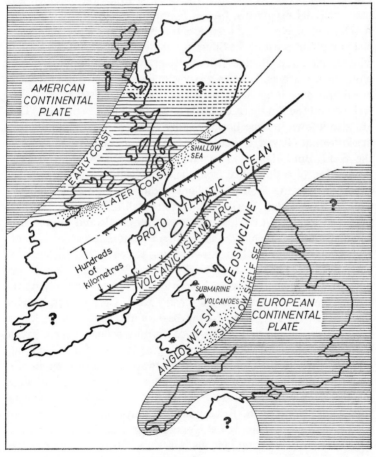

FIG. 15
Reconstruction of the supposed geography of Ordovician times.

insignificance, but the geosynclines still persisted along their former lines. As in the Ordovician, graptolites (of new kinds) were the most useful animals to be fossilized, accompanied by a great variety of new brachiopods and trilobites. Many other types of fossils appear for the first time in abundance in Silurian rocks. The molluscs, for example, by now included many **bivalves** (comparable to the modern cockle and oyster) and the more or less coiled **cephalopods** (which were to become far more important later and were to leave such modern descendants as the cuttle-fish and the pearly nautilus). **Corals** also became numerous for the first time and together with other colonial organisms, form a major constituent of many rocks. A selection of typical Silurian fossils is shown in Figure 17.

Throughout most of the Silurian, the thick deposits of the geosynclines remain quite distinct from those of the very shallow 'shelf' seas along their margins. Silurian geosynclinal sediments with graptolites are found over large areas of Wales, the Lake District, north-east Ireland and the Southern Uplands of Scotland (see Figure 13). They are also known in deep borings under Kent. The marginal deposits are known at Girvan, as before, and in a broad belt extending down the Welsh Borderland and round into South Wales as far as Pembrokeshire. One difference from the Ordovician was that a shelf sea extended over the English Midlands (compare Figures 15 and 19). Thus shallow-water Silurian deposits are well-known near Birmingham, for example in the hills where Dudley Castle and the zoo now stand. South of Birmingham, in the Bristol Road near the old Rubery mental hospital, is a fine section which clearly displays the early history of the area. A bright red sandstone of Silurian age, with its characteristic fossils, is seen resting unconformably on the white Cambrian quartzite of the Lickey Hills. The junction between the two is very irregular and the base of the sandstone is full of boulders of the quartzite. This unconformity accounts for a gap in the geological record which covers the greater part of the Cambrian, and the whole of the Ordovician. It is always difficult to say, in such circumstances, if strata of these ages were never deposited at all, or whether they were worn away before the later strata were laid on top. At one point in the Rubery section, a dyke can be seen cutting the Cambrian quartzite, but stopping abruptly at the unconformity. This is one of the many igneous intrusions already mentioned (see page 65) which are thought to have been emplaced in Ordovician times.

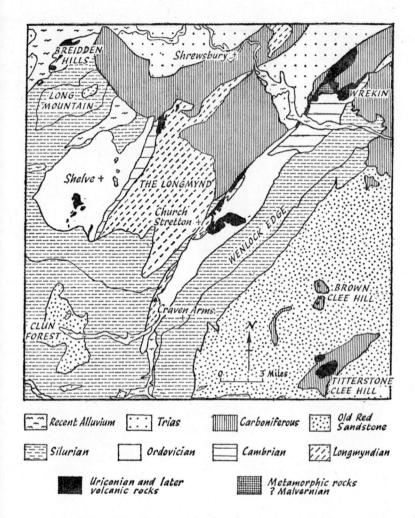

FIG. 16

Geological sketch-map of South Shropshire (based on published maps of the Institute of Geological Sciences and other sources).

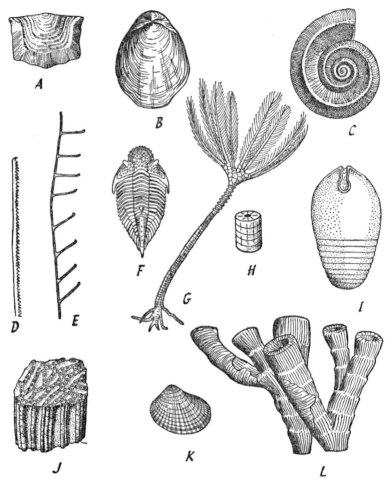

FIG. 17

Silurian fossils. *A* and *B*: brachiopods, both × ⅔; *C*: marine snail or gastro-pod, × ½; *D* and *E*: graptolites, *D* × 1, *E* × ¾; *F*: trilobite, × ⅔; *G*: crinoid or 'sea-lily' showing 'roots', stem, cup and arms, × ½; *H*: segments of crinoid stem, × 1; *I*: aberrant nautiloid mollusc (after Zittel), × ⅔; *J*: compound coral, × ⅔; *K*: bivalved mollusc or lamellibranch, × ⅔; *L*: branching compound coral, × 1 (*A*, *D*, *E*, *F*, *J* and *K* after Institute of Geological Sciences figures).

The Silurian shelf-sea deposits are very distinctive and uniform. They are perhaps best studied in east Shropshire. This is classic ground for the geologist, and one can hardly walk anywhere here at certain times of the year without meeting parties of students or individuals industriously tapping at the rocks with their hammers. Shropshire place-names such as Wenlock and Ludlow are used all over the world for divisions of the Silurian System, and Sir Roderick Murchison first named this major part of geological time after the ancient British tribe—the Silures—who lived in this region (Figure 16).

In the centre of Shropshire, west of Church Stretton, is the upland area of the Longmynd, composed of Precambrian sedimentary rocks. Round the southern edge of the Longmynd are the remains of shingle beaches left by the first Silurian sea which swept across the area. Many shoreline features can be traced—bays and inlets, sea-stacks and shingle-spits. It seems that the Longmynd stood up as an island for a while, before being buried under later sediments. Erosion has now just reached the right level to reveal these ancient features once more.

Just east of Church Stretton is the small but impressive mountain of Caer Caradoc. This is one of the many hills in the Welsh border-land where Caractacus is said to have made his famous last stand against the Romans. Caradoc is the highest of a line of whale-backed hills formed of Precambrian volcanic rocks. The one immediately to the north is the Lawley, and from here a pleasant day's walk east-wards takes one across the whole of the Lower Palaeozoic succession (see Figure 18). The walk illustrates very clearly the way in which the nature of the rocks determines the nature of the ground and its vegetation.

The formations are all sloping down towards the south-east,

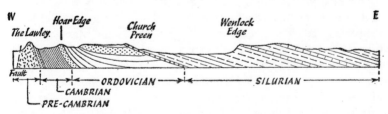

FIG. 18

Cross-section of the Lower Palaeozoic country in East Shropshire (from a section on the Shrewsbury map of the Institute of Geological Sciences).

rather like a line of books which have toppled over. The softer formations form valleys and the harder ones ridges. Since the **dip** (that is the downward slope of the beds) is gentle, the easterly or 'dip slopes' are also gentle. The westerly slopes (**'scarp** slopes' or 'escarpments') on the other hand, are steep, and one is soon aware of the nature of the formations one is crossing.

The basal Cambrian quartzite is plastered on the eastern face of the Lawley and the remainder of the Cambrian, being fairly soft sediments, forms low-lying ground. There are two sandstones in the Ordovician which forms marked escarpments, then, as one descends the second dip-slope, one comes on to the Silurian. This system has also produced two escarpments here, in this case formed by limestones, and it is the first of these which is the highlight of the day's walk. This is the fine, wooded feature of Wenlock Edge, which dominates Shropshire scenery. It is well known in poetry, prose and music, but to the geologist it means one thing above all—the Wenlock Limestone. This limestone comes about the middle of the Silurian and is famous because of the extreme abundance and variety of its fossils. It is a good example of an 'organic limestone', being largely made up of the remains of animals with limy skeletons or shells. Other limestones are composed of lime compounds which were formed as a chemical precipitate in the water. Almost every slab of Wenlock Limestone is packed with brachiopods, corals, cephalopods and members of other groups. **Crinoids** are also particularly abundant. These were static, flower-like sea-creatures, popularly known as 'sea-lilies' (see Figure 17). Many limestones are almost entirely composed of disassociated segments of crinoid stems. Trilobites are common in some places and rare in others. They are so abundant in the old lime-workings at Dudley that one is incorporated in the town's coat of arms.

One especially interesting feature of the limestone is the presence in it of reefs, comparable to, though smaller than, the coral reefs of the present day. These reefs are built up of corals and other colonial organisms, together with simple plants, all more or less in the position and attitude in which they lived. They stand out as ball or mushroom-shaped masses of unbedded limestone, surrounded by normal layered limestone. One can picture a warm, shallow sea extending over the English Midlands in middle Silurian times, teeming with marine invertebrate life of all kinds and dotted with reefs. The Niagaran limestones of North America, which form the strong, precipitous feature which gives rise to the Niagara Falls, are about the same age as our Wenlock Limestone and are equally fossiliferous.

Though marine life was so abundant where the Wenlock Limestone was being deposited, it should be remembered that the plants were still all of the simplest kinds (algae) and the animals were still all invertebrates—that is without backbones. The skeletons which have been mentioned so far have all been limy external skeletons like those of crabs or lobsters; there were no animals with bony internal skeletons in the Wenlock sea. In other words, there were no fish in the sea and there were no plants on land. The oldest known land plant, with all the complicated structures necessary for terrestrial life, was found in the Upper Silurian of Australia. The oldest known fish comes from a Middle Ordovician sandstone in Colorado. Though this last sentence may seem to contradict what went before, it does not in fact do so, for the sandstone in question was laid down in freshwater and it was not until much later that fish entered the sea. Fish do not otherwise appear in the record until right at the top of the Silurian.

The arrival of fish in Britain is marked by a remarkable, though very thin, deposit. This is called the 'Ludlow Bone Bed' and is well known along the borders of England and Wales, where it forms a useful horizon for the geologist when he is making a geological map. It is usually no more than two or three inches thick and is packed with the scales, bones and teeth of primitive fish. When fossils were first studied, it was realized that they differed from one formation to the next, and it was thought that the life of one age was destroyed in a great catastrophe before being replaced by an entirely new creation. This doctrine was only rejected after many years of argument, partly because of the new ideas of evolution, and partly because of increasing knowledge which revealed more and more faunas and therefore required more and more catastrophes until the whole concept became ridiculous. The Ludlow Bone Bed was thought to provide a clear demonstration of one of these catastrophes in which all the life in the sea at the time was suddenly destroyed. Nowadays it is thought that such deposits were not the result of a single great disaster, but of a long period of very slow deposition, when the sea-floor was being swept by strong currents. These carried away the lighter sediment and left the heavier fragments, such as bones, to accumulate. In any case the lifetime of an organism is such a fleeting thing and sedimentation is usually so slow, that all the organisms living in the sea at any one

FIG. 19

Reconstruction of the supposed geography of Silurian times (after Wills and others).

moment must be dead before more than a microscopic amount of sediment has been deposited.

It is thought that the fish remains of the Ludlow Bone Bed were freshwater creatures brought down by rivers into the sea where the bed was formed. From that level onwards, there is a more or less complete record of fish, leading to amphibians, to reptiles and thence to mammals. The explanation for this sudden appearance of vertebrate animals seems to be that they first evolved in freshwater, in rivers, and lakes, though presumably they must originally have derived from sea animals. Most of our Lower Palaeozoic rocks were laid down in the sea, and it is not until we get a great change-over to land conditions at the end of the Silurian that fish appear in our story. At the same time we get the first higher plants, adapted for life on dry land.

The land which emerged in the British area in the Devonian Period is described in Chapter V, but linked with this important change was an interlude of great upheaval and disturbance which is dealt with in the next chapter.

IV

THE GREAT EARTH MOVEMENTS

This chapter deals with the mountain-building earth-movements which occurred at the end of the Lower Palaeozoic, and which are known as the Caledonian Orogeny. At the same time it introduces some of the principles of structural geology

---

The cores of the continents are the great shield areas of Precambrian rocks which were discussed in Chapter II. These ancient rocks themselves had very complicated histories, with repeated earth-movements going far back beyond our knowledge, but we are only concerned with them now as the nuclei round which sediments accumulated and against which these sediments were piled in subsequent mountain-building movements. The greatest thicknesses of sediment always accumulated in geosynclines, like those described in the last chapter. These geosynclines were more or less marginal to the great shields. Thus the Canadian Shield with its platform extension under the central U.S.A., had, in Lower Palaeozoic times, the Appalachian geosyncline on its eastern margin and the Cordilleran geosyncline on its west.

It is the fate of every geosyncline to become a mountain range. Up and down the stratigraphical column all over the world, there are examples of geosynclines—great troughs full of sediment which accumulated slowly through many millions of years. In every case, at the end of the long period of sedimentation, there were paroxysmal movements which heaved up the thick pile of sediment as new mountain chains. The Alps and the Himalayas, the Rockies and the Andes, are all mountain ranges produced in this way comparatively recently in geological time. The Appalachians are an example of a much older mountain chain, now worn down by long ages of erosion. In fact, just as 'every valley shall be exalted', so shall 'every mountain and hill (be) made low', and the time required for each process is probably of about the same order.

In Britain, the geosynclines which had been filling with sediment all through Lower Palaeozoic times, were crumpled up to form mountains in what is now Wales, the Lake District and the Southern Uplands of Scotland (see Figure 20). These areas are still mountainous today, but the original mountains were worn away long ago, and the present ones are of comparatively recent origin, standing up as they do because of the hardness of their ancient rocks.

The cause of such mountain-building movements or **orogenies** is now much more clearly understood in the light of plate tectonics. Where a spreading oceanic plate meets a continental plate that is either static or moving in the opposite direction, something violent is bound to happen. The heavier oceanic plate tends to be pushed down beneath the continental plate in a subduction zone under conditions of great pressure and comparatively low temperature.

The term 'geosyncline' has been applied to a variety of sedimentary basins, but in the strict sense it can be said to correspond to the long oceanic trenches that form by the down-drag of subduction zones (see Figure 4). So we now interpret the Lower Palaeozoic geosynclines of the British Isles as the oceanic trenches on either side of an early version of the Atlantic Ocean.

The tremendous strains and pressures set up by the collision of two plates leads firstly to the over-riding of the ocean by the continent and secondly to the intense crumpling of the lighter continental material (Figure 5). The oceanic trench or geosyncline begins to fill with sediment as the oceanic plate is forced downwards and then these sediments become involved in the compression and are pushed up to form part of the continent. These are the thick sandy mudstones that form so much of the Lower Palaeozoic succession—for example the regular layers in the cliffs around Aberystwyth.

Usually it is not the over-simplified picture shown in Figure 4, but parts of the oceanic plate itself get pushed up into the new mountains by the process known as obduction (Figure 5). So along the coast south of Girvan in south-west Scotland one finds the pillow-lavas and the thin deep-sea sediments that testify to the presence of oceanic crust. We see them too in the Lake District and from Snowdon down to Strumble Head in Pembrokeshire. Also under the conditions of great pressure, the oceanic crust is altered and masses of 'greenstone' or serpentine originating from the peridotite layer (see Figure 5), are emplaced amongst the chewed-up sediments and lavas of the ocean floor. This is derived from the lower peridotite layer of the oceanic crust.

The earth-movements which occurred in north-west Europe at the end of the Lower Palaeozoic are known as the **Caledonian Orogeny**, after Caledonia or Scotland where their effects are particularly well shown. By the end of Silurian times, sedimentation had caught up with subsidence in the British geosynclines, so that conditions were more or less uniform throughout the areas concerned. The great heap of sediment was continuously being involved in the driving together of the two plates and so the edge of the continental plate became thickened. As this material was relatively light (compared with the heavy rocks of the oceanic plate and the mantle down below) it tended to rise. This elevation in turn probably produced further sliding movements under the influence of gravity. It also produced, by erosion, more sediment to pour into the neighbouring geosynclines, which continued—as it were—to feed on themselves.

Eventually, at about the end of the Silurian, conditions became stabilized, the subduction zones stopped subducting (or should it be subducing?) and the geosynclines or oceanic troughs ceased to exist. The Moffat geosyncline probably ended much earlier (see Figure 19).

However, the process was not finished. The sinking of the heavy oceanic crust below the continents took it into conditions of great pressure and high temperatures due mainly to friction, with consequent melting. The resultant molten rock forced its way to the surface as volcanic effusions of a different type from those in the middle of oceans, in particular the lavas known as andesites. Also the formation of mountain ranges in the lighter (less basic) continental crust produced metamorphism and the emplacement of granitic masses, partly from the melting of the 'roots' of the mountains as they sank into the deeper, hotter regions of the interior.

In this way geosynclines are thought to perpetuate themselves for many millions of years. At the end of their history, however, the up and down movements give way to convergent lateral movements and the great heap of sediment is crumpled together and so made even thicker. The mountain ranges are produced not only by the actual crumpling, but also by the rising of the vast mass of lighter material, like an iceberg in water.

The Caledonian Orogeny—like all orogenies—affected the rocks in various ways. When rocks are subjected to lateral pressure, they

GRANITE

Skiddaw
Shap

Map showing the main trends of Caledonian folding in Britain, with some of the granite masses emplaced at this time.

at first tend to bend like a sheet of modelling clay. In other words, they deform plastically into **folds**, and this is a suitable point to discuss these widespread geological phenomena. When rocks are folded upwards into an arch, this is called an **anticline** (see Figure 21A), when they are folded downwards, this is a **syncline** (Figure 21B). These are the simplest types of folds and may differ considerably in magnitude. The Wealden area (see Figure 46) is an example

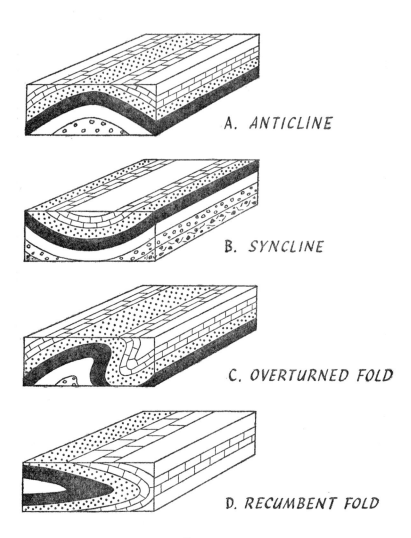

A. ANTICLINE

B. SYNCLINE

C. OVERTURNED FOLD

D. RECUMBENT FOLD

Fig. 21
Different types of fold.

of a very large anticline which is rounded off at the ends in the form of an elongated **dome**. Other anticlines have only an amplitude of a few feet. Areas such as the Southern Uplands of Scotland have more complex folds produced in the Caledonian Orogeny. Many of the anticlines are **overturned** (Figure 21c) in which one side or 'limb' has been pushed beyond the vertical. This may go a stage farther and the fold may be lying on its side or **recumbent** (Figure 21D).

In the Precambrian rocks of Scotland, folding is even more intense (as shown by Plate 2) and it can often be shown to have occurred several times along different lines. More complicated structures will also be discussed in connection with the Alps (see Chapter X).

Naturally, a substance like a bed of rock can only bend to a limited degree before breaking. A rock such as shale will fold very easily under pressure but harder sedimentary rocks and volcanic rocks very soon fracture. So, in addition to the folding produced in the Caledonian Orogeny, there was also extensive fracturing or 'faulting'. A **fault** is any abrupt displacement of beds produced by breaking and not by bending. There are essentially three types of fault; (i) **normal faults**, produced by tension (see Figure 22A); (ii) **reversed faults**, produced by lateral pressure (Figure 22B); (iii) **tear faults** produced by lateral torsion (Figure 22c). These last may be difficult to recognize on the ground, because the strata may appear to be the same on either side. However, in the simplest case, there may be an igneous intrusion which clearly reveals the lateral displacement, as shown in the diagram. Such an intrusion was used by Professor W. Q. Kennedy in demonstrating the most remarkable tear-fault produced in the Caledonian movements. This is the 'Great Glen Fault', which runs right across the Highlands from the Moray Firth to the Firth of Lorne along a line of magnificent lochs (see Figure 23). Professor Kennedy showed that the Strontian Granite, on the mainland opposite Mull, was probably a continuation of the Foyers Granite south-east of Loch Ness. If this is true, it means that the two granite masses were torn apart in the faulting and that the whole of northern Scotland moved about 65 miles from north-east to south-west as shown in Figure 23. A thrust which passes through the island of Islay is probably the continuation of the Moine Thrust which is discussed below. This is a very unusual tear fault, but there are others like it in other parts of the world. Perhaps the most famous is the San Andreas fault in California, which moved a little

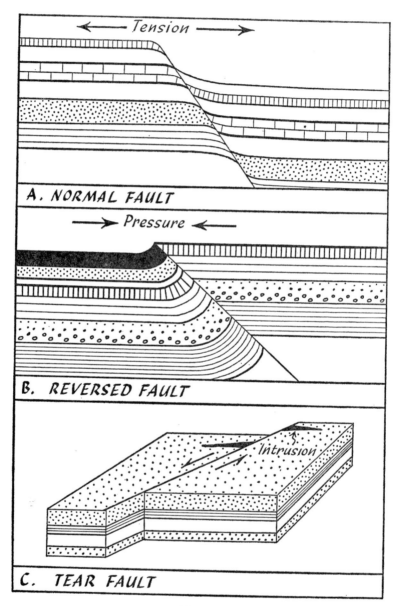

Fig. 22
Different types of fault.

in 1906 to cause the disastrous San Francisco earthquake of that year.

As a result of the great lateral pressures operating during an orogeny, reversed faults are commonly associated with the intense folding described above. Reversed faults which are inclined at an angle of less than 45° to the horizontal are known as **thrusts**, and there are many such, in Scotland and Wales particularly, which can be blamed on the Caledonian Orogeny. The best known and most spectacular thrust in these islands is the Moine Thrust which runs roughly parallel to the north-west coast of Scotland for many miles (see Figure 23). It must have occurred before the Great Glen Fault, since it is itself affected by that structure.

In this thrust (or rather this series of thrusts) a great mass of Moinian rocks was pushed over the Lewisian, Torridonian and Cambrian of the west coast. In the confusion which resulted, the various formations are found lying one upon another in a complex heap of thrust slices, so that in places Precambrian rocks are seen above the Cambrian. Such anomalies produced much of the early uncertainty and controversy about the geology of this part of Britain.

The great pressures associated with the thrusting and the grinding together of the rocks produced local metamorphism which further obscured the record. It may be that the main regional metamorphism of the Highlands occurred during the Caledonian Orogeny. The Lewisian rocks had been metamorphosed long before, but the great areas of Moinian and Dalradian rocks, described in Chapter II, may have undergone their chief alteration at this time. Near Bergen in Norway there are schists containing Lower Palaeozoic corals and molluscs. They are distorted and altered, but identifiable, and prove that many of the metamorphic rocks of Norway may be of Lower Palaeozoic age. By the same argument it has been suggested that the Dalradian rocks of Scotland, which form a natural continuation of the Norwegian rocks, may also be metamorphosed Lower Palaeozoic sediments.

The Moinian and Dalradian rocks of the Highlands would be regarded as the deep-seated roots of the mountains that were formed during the orogeny, but have long since been eroded away. Regional metamorphism must have occurred at a considerable depth below the surface. In areas such as North Wales, however, though the Lower Cambrian sediments must have been buried beneath nearly seven miles of rock, they have hardly been altered at all.

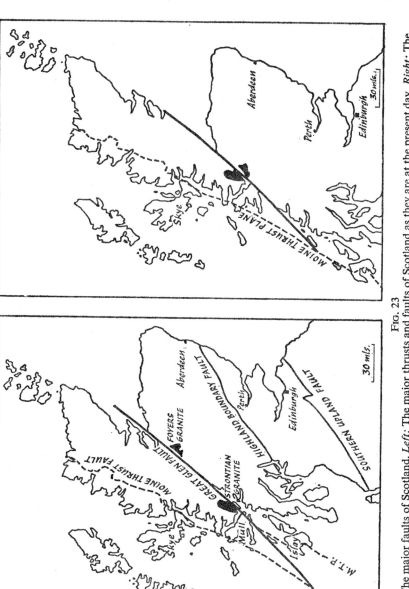

Fig. 23

The major faults of Scotland. *Left*: The major thrusts and faults of Scotland as they are at the present day. The position of the present land areas as they are thought to have been before the movement of the Great Glen fault. Note the position of the two granite masses and the displacement of the Moine Thrust plane (based on Professor Kennedy's maps, by kind permission of the author and the Council of the Institute of Geological Sciences). There is now evidence

THE GREAT EARTH MOVEMENTS

The chief metamorphic effect of the Caledonian Orogeny in Wales, as in the Lake District and the Southern Uplands of Scotland, was the production of **slates**. Slates are muddy sediments which under great pressure have developed a **cleavage** along which the rock readily splits. The cleavage is produced by the platy minerals in the rock becoming reoriented at right angles to the direction of greatest pressure, and it often cuts across the original bedding. The cleavage is therefore related not to the way the rocks were originally deposited, but to what happened to them afterwards. It is therefore often a useful guide to the structural history of a region. Cleavages can also be formed in other ways.

Another feature of the orogeny in all the areas affected, was the emplacement of large and small igneous intrusions. There are, for example, many large granite masses in the British Isles which probably reached their present position at this time. There were a whole series of granites emplaced in the Scottish Highlands, notably the famous ones around Aberdeen, which have been quarried on a huge scale for building stone. The Lake District has a series of smaller granites, the best known being the one that forms Shap Fell. This is commonly used as an ornamental facing-stone and can be easily recognized by its large pink felspar crystals.

Since the deep trough of Cambrian to Silurian sediments was orientated in a north-east to south-west direction, this was also the main trend of the new mountain ranges. A glance at an ordinary map of Scotland is sufficient to see the 'grain' imparted to the region by the Caledonian movements, and it shows even more clearly on a geological map. Major features such as the Great Glen, the main trend of the Southern Uplands and the whole shape of the country tend to reflect the old mountain ranges. The same north-east/south-west lines are seen in the older rocks of Ireland, in the Lake District, in Central Wales and in the Welsh Borderland (see Figure 20). Some movements happened much later along the same lines. Thus, running from Shropshire down into South Wales is the Church Stretton fault which affects rocks as young as the Triassic, but which was obviously an important line as far back as the Ordovician.

THE OROGENY AND EVOLUTION

A subject which has long been a matter of argument and discussion among geologists is the relationship between major earth-movements, such as the Caledonian Orogeny, and organic evolution. When important changes in the animal and plant kingdoms coincide with orogenies, it is tempting to suggest a direct connection. Thus, at about the time of the Caledonian movements, the graptolites became extinct, the trilobites showed a marked decline, the land plants and the coiled cephalopods known as 'goniatites' appeared for the first time and vertebrates suddenly came into the record in force. There were many other similar changes, but it is difficult to be sure that these occurred all over the world at the same time, and not just in the areas which have been most studied. No one would nowadays suggest that major groups of animals or plants were wiped out in great catastrophes, but there may have been a more indirect control of the organic world by inorganic phenomena.

A period of great earth-movements such as the Caledonian Orogeny must have considerably altered the geography of the world at the time. Seas became land, climates changed from humid to arid and mountain ranges appeared where they had not been before. The organic population of a particular place had to move, change or perish. The permanent and complete migration of a species into another area is rarely easy. The area may not be suitable and is likely to be already occupied by rival species. What is more, the changed geography may give new opportunities to groups previously living in obscurity. Every organism has certain intrinsic potentialities in its make-up which are not realized until circumstances make it possible. Under stable conditions, the processes of natural selection will favour conservatism; they will favour the organism which is perfectly suited to the environment of its ancestors. But under changing conditions, natural selection will favour the novelties, which may not be suited to their ancestors' environment, but may be better equipped than any of their contemporaries for the new situation brought about by geographical changes. In other words, evolution is very much a matter of opportunity.

The most obvious example of this which may have been brought about by the Caledonian Orogeny was the rapid spread of land plants. The sudden appearance of vast areas of dry land where previously there had been sea, provided an opportunity for the ancestral land plants to spread at a remarkable rate, colonizing the new environment.

One must always be sceptical, however, of such easy generalizations. It is very difficult to be sure that fossil changes were as sudden as they seem. Orogenic periods are times when the record is least well preserved and most confused, so that the very different faunas before and after an orogeny may only be an indication of the length of the gap.

THE AGE OF THE CALEDONIAN OROGENY

It would seem to be implied by the placing of this chapter, and by statements already made, that the Caledonian Orogeny occurred in that infinitesimal moment of time between the Silurian and the Devonian Periods. This, of course, cannot be true, but it is the sort of impression that must be given by any book which tries to reduce a number of complex, contemporaneous processes to a few simple statements. We say that the Caledonian Orogeny occurred at the end of the Silurian because sediments which were laid down up to that date tend to be affected by it, whilst later ones do not. But there are places—in south Shropshire for example—where gently dipping Silurian strata pass up gradually into Devonian strata with no signs of a break or any sort of disturbance. There are also places where important earth movements occurred much earlier, and others where they went on during the Devonian. Some of the intrusions in the southern Highlands were particularly late.

The idea of quite sudden and violent mountain-building movements probably derived from the older ideas of catastrophism. The majority of geologists now believe that orogenies were very lengthy affairs, which built up slowly to a climax and then died away equally slowly. The most that we can say about the age of the Caledonian movements is that they probably reached their maximum intensity at about the end of the Lower Palaeozoic. They may have started as early as the Cambrian, where there are local unconformities, and there is evidence of important movements during the Ordovician in the Lake District and on the Ayrshire coast. It is now thought that the main metamorphism in the Highlands might be mid-Ordovician in age.

In North America, the main orogeny of the Lower Palaeozoic occurred at the end of the Ordovician, and movements in Britain at this time can be demonstrated in Shropshire and elsewhere. Still, the Southern Uplands clearly suffered their intense folding in late Silurian times and this also fits the evidence in many other areas. Perhaps the best evidence of all lies in the great rush of coarse sediments from nearby mountains which appeared early in the Devonian record. This and the diminuendo of the Caledonian Orogeny will be discussed in the next chapter.

V

DRY LAND AND SHALLOW SEAS

This chapter concerns the Devonian and Carboniferous Periods, which followed the Caledonian Orogeny and preceded the Hercynian Orogeny at the end of the Palaeozoic. They were times of dominantly shallow marine and terrestrial deposition.

THE DEVONIAN PERIOD

The landscape which emerged at the end of the Caledonian Orogeny was quite different from what had been there before. For many millions of years, the area which is now north-west Europe had been dominated by great marine geosynclines. Now came the emergence of an extensive landmass with mountain chains running from south-west to north-east, from southern Ireland to northern Norway. The new mountains were immediately attacked by the elements, and the detritus so produced accumulated in the valleys. Tremendous thicknesses of sediments were built up on dry land instead of in the sea, and for the first time since life appeared on the earth, we have a good record of land organisms.

In Chapter III, the British stratigraphical record was traced up as far as the Ludlow Bone Bed. At that level fish remains are found for the first time, and this is attributed, not to the sudden evolution of fish, but to the onset of continental conditions. The supposedly fresh-water fish are of a very simple type, without proper jaws, and encased in a heavy external skeleton (see Figure 24). They are comparable in many features with the modern lamprey, to which they are distantly related; but this is one of many cases in biology in which the modern representatives provide only a poor and probably misleading guide to a group that is largely extinct. The jaw-less fish became all but extinct before the end of the Devonian Period. The few living

91

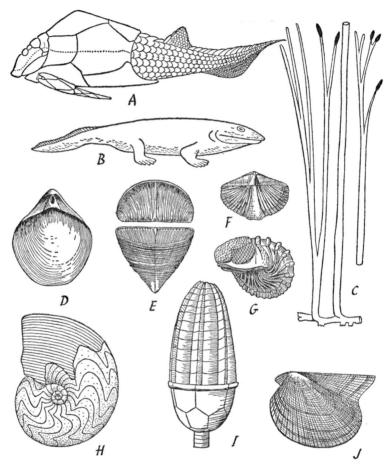

FIG. 24

Devonian fossils. *A*: primitive, heavily-armoured fish (after Watson), × ½;
B: reconstruction of a primitive amphibian showing fish-like tail (this form may
possibly be early Carboniferous in age), × 1/16; *C*: primitive land plants from the
Rhynie Chert (after Kidston and Lang), × ⅓; *D* and *F*: brachiopods from the
marine facies, both × ½; *E*: simple coral with lid, × ½; *G*: trilobite in rolled-up
position, × ⅔; *H*: early ammonoid or goniatite with shell worn away showing
internal partitions, × ⅔; *I*: cup and arms of a crinoid, × ¾; *J*: bivalved mollusc
or lamellibranch, × ⅔.

representatives have therefore had nearly 300 million years in which to change and become less and less like their ancient relations. The lamprey is a degenerate, eel-like, parasitic animal, but the Devonian stock from which it came must have been progressive and far from degenerate, since it probably gave rise to all the later groups of bony animals.

An interesting feature of the earliest fish is that they carried most of their bone externally. They were mostly heavy, sluggish creatures which moved about on or near the bottom, probably shovelling up food and mud in a scoop-like mouth. Early in the Devonian, there appeared the next stage in fish evolution, which were forms having simple hinged jaws and paired fins, but these passed their peak within the Devonian and were extinct before the end of the Palaeozoic. By the end of the Devonian, the remaining two major groups of fish had also appeared—the cartilaginous fish such as the sharks (which have no bony skeleton) and the true bony fish with fully-evolved jaws, which dominate the seas at the present day. Apart from the biological interest of their evolution, the Devonian fish are important to the geologist because they are almost the only fossils in the thick continental deposits which can be used for correlation.

Undoubtedly the most important biological change which occurred at the end of the Lower Palaeozoic was the evolution of land plants. Until there was land vegetation to provide food, there could be no land animals, and life was restricted to the water. Plant remains are the oldest known fossils, way back in the Precambrian, and similar forms are found all through the Lower Palaeozoic, but all of these are **algae**—plants of the simplest type—with the habit of secreting about themselves layers of calcium carbonate. Seaweeds or 'brown algae' were probably also abundant, but had no hard layers to fossilize. All such plants depend on the water to buoy them up; they take in their necessary gases from the water and their reproductive spores are distributed by the water. For a plant to live on dry land it has to have some sort of stiffening tissue to hold it up, it must be able to take in gases from the air and it must be able to distribute its spores through the air.

Shortly before the First World War a Scottish geologist, Dr. Mackie, was studying the Devonian volcanic rocks near Rhynie in Aberdeenshire. He collected some specimens of what he took to be a volcanic lava flow. When his specimens were cut into thin slices for examination under a microscope, he was astonished to find not the

FIG. 25
Map showing the main outcrops of Devonian rocks in Britain.

glassy structures of a lava, but the perfectly preserved tissues of land plants. The Rhynie plants had apparently been living in a rather damp peat bog which was soaked with mineral-rich waters—perhaps from local volcanic hot springs. The peat was petrified, being turned into a rock of almost pure silica—known as **chert**—in which the delicate tissues of the plants were completely preserved. They were undoubtedly land plants, though of a very primitive type. They had central strands of woody tissue to hold them erect, they had the pores through which gases pass in and out of land plants, and they had special devices for scattering their spores in the air for dispersal by the wind. On the other hand, they had no true roots for taking in water and minerals from the soil, they had no true leaves and their woody tissues were of the simplest kind (see Figure 24).

The Rhynie Chert is Middle Devonian in age or perhaps a little older. Other plants are known in Devonian rocks, but none so well preserved. The earliest known of all land plants was found in Upper Silurian sediments in South Australia, but the main record started with the emergence of great areas of dry land in the Devonian. By the end of that period most of the main groups of plants had evolved, excepting only the most advanced of all—the flowering plants.

The whole trend of evolution in the plants seems to have been an increasing adaptation to dry conditions. The Rhynie plants were land forms, but still required the damp habitat of a peat-bog. Before the end of the Palaeozoic, plants had colonized the dryer upland areas and have evolved the seed, in which the reproductive structures are completely encased and protected from desiccation.

The fish and plants of the British Devonian are found in the thick continental sediments which have been known since the earliest days of geology as the **Old Red Sandstone**. This name is perhaps not altogether suitable, for the sediments are not always red and they are often other than sandstones. But, for general purposes, the name does give the right impression of the coarse oxidized sediments which accumulated rapidly in the hollows between the new mountains. The Old Red Sandstone is by no means evenly distributed. It is very thick in one place and very thin, or altogether absent, not far away. Thus it is thousands of feet thick in South Wales and in the neighbouring counties of England, but it is completely missing in the English Midlands. Apparently it formed in a series of inter-montaine basins scattered across northern Europe. In Great Britain there were three main basins, one in the north of Scotland (including what

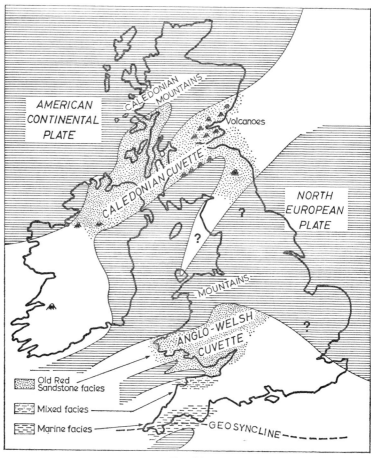

FIG. 26

Reconstruction of the supposed geography of early Devonian times (simplified from *A Palaeogeographical Atlas of the British Isles* ... by L. J. Wills, by kind permission of the author and Messrs. Blackie & Son, Ltd.). The north Scotland or *Orcadian* cuvette did not come into existence until later.

are now the Orkneys), a second in the south of Scotland and a third included South Wales and adjacent parts of England (see Figure 26).

The Old Red Sandstone can be subdivided, chiefly by means of its fish faunas, into three series—the Lower, Middle and Upper 'O.R.S.'. These are not all present everywhere, and the middle division in particular is usually missing. This must have been the result of late episodes in the Caledonian Orogeny. In places, both the lower and

96

the middle divisions are missing. At Siccar Point, south of Dunbar on the coast of Berwickshire, the Upper Old Red Sandstone is seen resting almost horizontally on sharply truncated, vertical Silurian slates. James Hutton described the section here in his *Theory of the Earth*, first published in 1795. He used the evidence of this section and others to demonstrate a 'succession of former worlds'. He was the first man who really understood the nature of sedimentary and igneous rocks and used field evidence to prove what had previously only been argued about in studies and libraries.

The Lower Old Red Sandstone is well developed in South Wales, especially in south Pembrokeshire where there are many thousands of feet of greenish sandstones and coarse pebble-beds. They are seen all round the great natural harbour of Milford Haven and extend round the north side of the South Wales coalfield into Herefordshire. A very large area along the borders of Wales is occupied by the Lower O.R.S. (see Figure 25). It extends from just north of Cardiff to near Much Wenlock in Shropshire, and from the Malvern Hills deep into Central Wales. In this region it is usually a dark red or reddish-brown sandstone and has been much used for building.

In this, as in many areas, there are thin-bedded 'flaggy' divisions; surfaces in these often show ripple-marks like those left on sandy beaches by the tide. They are a sure sign of shallow water deposition. Sometimes the surfaces are covered with hexagonal cracks which show up because they are filled with sediment of a slightly different colour (see Plate 4). These were produced when the soft sediment was exposed for a while to the hot sun; as it dried, it contracted and cracked like the mud in a modern dried-up pool.

The Upper O.R.S. rests directly on the lower division throughout South Wales, with no sign of intermediate formations. In places there is an obvious unconformity between the two, and the higher beds contain breccias and conglomerates. Apparently deposition stopped for a long time in this region before renewed uplift of the mountains brought fresh supplies of coarse sediment pouring into the basin. Variously-coloured sandstones and conglomerates of the Upper Old Red Sandstone form a capping to some of the finest local scenery. The rocks dip southwards under the coalfield and so produce impressive north-facing escarpments such as the Brecon Beacons and the Black Mountains.

A single tiny patch of Upper Old Red Sandstone is found in England—on Titterstone Clee Hill in Shropshire. This indicates the former wide extent of the formation, most of which has now been eroded. Lower Devonian continental rocks appear again in southern Scotland and in adjacent parts of north-east England. Here the Old Red Sandstone differs from that of the south in that it contains great thicknesses of volcanic rocks interbedded with the sandstones and conglomerates. In places in the Midland Valley of Scotland there are nearly 20,000 feet of the Lower O.R.S. alone. The volcanic rocks include lavas of all types, ashes and volcanic boulder beds or **agglomerates**, besides intrusive dykes and sills (see Plate 5). They are well seen in the Pentland Hills south of Edinburgh. In several places the actual vents of Devonian volcanoes can be recognized. These are filled with hard agglomerate and so have survived whilst the soft ash cones which surrounded them have been worn away. Mochram Hill near Maybole in Ayrshire is an example of such an eroded volcano.

Associated with the volcanic rocks are many interesting minerals, though few of economic importance. Lava-flows are often full of gas-cavities. These are usually filled later with special minerals of varied composition, notably beautifully colour-banded agate and onyx. These are harder than the lavas, and so weather out to be found as loose pebbles in rivers and on beaches. They are polished and used as semi-precious stones under the popular name 'Scotch Pebbles'.

Just south of the border in the north-east corner of England is the old cattle-raiding country of the Cheviots. These hills represent a deeply eroded volcano of considerable size. The volcano started with a series of great explosions which scattered quantities of ash and agglomerate over the surrounding area. Next there were extensive outpourings of lava, through which a large plug of granite was pushed up from below. Finally a number of dykes were implaced in both granite and lavas, radiating from the central point of weakness which produced the volcanicity. They may be compared with the cracks radiating from a bullet-hole in a window.

During Middle Devonian times in the south of Scotland there was considerable uplift, folding and erosion. The Upper O.R.S. once covered a very large area. In places it is seen resting unconformably on the lower division. Elsewhere (as at Siccar Point) it rests on vertical Silurian strata and away to the north it rests directly on Dalradian schists. There is evidence that some of the Upper O.R.S. accumulated under desert conditions. The sand-grains are perfectly rounded as they are by the wind in modern deserts (water has a cushioning effect and the grains do not rub together so much). The sandstones often show 'false-bedding' which seems to be of the same kind as that seen in sand-dunes today. It must not be thought, however, that desert conditions were normal in Devonian times. Apart from other evidence, the abundant plant-life makes this unlikely.

The Midland Valley of Scotland is bounded by two great faults which separate it from the other two main natural regions of Scotland—the Southern Uplands to the south and the Highlands to the north. These two faults appear to have been initiated in Lower Devonian times, though they moved again later. The northern or 'Highland Boundary Fault' still moves very slightly at times, producing tiny earth tremors. North of this fault, the Lower Old Red Sandstone is seen again in the volcanic area of the Lorne Plateau in Argyllshire, between Loch Awe and Loch Linnhe. A thin development of sandstone, shales and conglomerates with fish and other fossils, serves to date a great thickness of lavas and ashes with associated granitic intrusions. At Glen Coe, Glen Etive and Ben Nevis, cylindrical masses of the earth's crust are thought to have subsided into a chamber of molten granitic magma. The latter rose through the cracks and filled the cavities so produced. At Ben Nevis this happened twice, so that a later inner granite is found within an older outer granite.

Farther north again, the Upper Old Red Sandstone is seen on both sides of the Moray Firth, and in Caithness, the Orkneys and Shetlands. It is comparable to its southern developments and rests unconformably on the metamorphic rocks of the Highlands or on the Middle Old Red Sandstone. The latter is only to be found in this part of Britain and is particularly interesting. It underlies the little-known lowland area beyond the Highlands, which forms the greater part of Caithness. Flaggy sandstones predominate and were extensively quarried to provide the famous Caithness paving-stones before the regrettable introduction of artificial slabs. Fossil fish are locally abundant and a quarry near Achanarras still yields numerous specimens to geologists. Large areas of Middle Old Red Sandstone extend round the Moray Firth and down the Great Glen on both sides of Loch Ness. There are also scattered outliers to the south including the one around Rhynie in Aberdeenshire with its famous fossil plants.

The great pioneer of the Scottish Old Red Sandstone was Hugh

Miller, who was born near Cromarty on the Moray Firth in 1802 and spent the first fifteen years of his working life in the local quarries. In this work and in walks along the nearby sea-shore, he studied these then little-known rocks and collected their remarkable fossils. He later wrote about them in a vivid and powerful style which combines an extreme accuracy of scientific observation with a stout defence of the Established Church. His most famous geological work *The Old Red Sandstone* opens with an intriguing exhortation to young men not to attend Chartist meetings, but to read the Bible and to study geology. The foundations of geology, like those of most branches of sciences, were laid by wealthy men with time to spare for intellectual pastimes. Hugh Miller was a brilliant exception to this rule.

So far in this chapter, no mention has been made of the south-western peninsula of England where Devonian rocks are widely developed. Indeed, this may be regarded as the most important area of all, for the Devonian System takes its name from the county of Devon. This region has been left till last because the rocks here are very different from the Old Red Sandstone. The difference is mainly one of Devonian geography, for whereas the O.R.S. was laid down almost entirely on land, the Devonian rocks of Devon and Cornwall were mostly deposited in the sea. We have in this contrast one of the clearest demonstrations in our record of a lateral change in the rocks due to different environments (see Figure 26).

Between Barnstaple in North Devon and Minehead in Somerset, the Devonian succession displays an alternation of marine and continental conditions. There are representatives of the Lower, Middle and Upper Devonian with abundant marine fossils such as brachiopods, corals and trilobites (see Figure 24). Unfortunately the rocks of Devon and their fossils have suffered very badly in later earth-movements so that the area is not really a suitable one to give its name to a major division of geological time. The German Rhineland or New York State, with clear successions and perfectly-preserved fossils, would provide much better standards of reference.

In between the marine horizons in north Devon are beds such as those of Woolacombe Bay and Hangman Point, which in every way resemble the Old Red Sandstone of South Wales. Thus the shoreline oscillated from north to south during Devonian times. When it moved northwards, marine conditions extended into what is now Pembrokeshire and even spread as far as Brecknockshire. When the coastline moved south, Old Red Sandstone conditions invaded south-west England and typical fish faunas have been found as far south as Newquay in Cornwall.

On the south coast of Devon, the Devonian succession is practically all marine with a large and varied fauna. The fossils are best seen in the beautiful limestones of Torquay and Plymouth. Some of these are bright pink in colour and are packed with corals and other colonial organisms. They are sometimes polished and used for ornamental purposes in various public places. These limestones must have been laid down in a clear shallow sea. The abundance of corals suggests warm if not tropical conditions, though it is a little unscientific to assume that reef corals always required such a climate. Besides the corals, there are many new types of brachiopods and molluscs. The graptolites were all but extinct, and the trilobites very much reduced in numbers. Perhaps the most important innovation among the Devonian marine faunas was the appearance of the **goniatites**. These were cephalopod molluscs, distantly related to the modern octopus, but living in a coiled shell which was divided into chambers by sharply folded partitions (see Figure 24). Though not as spectacular as the evolution of the fish or land plants, the evolution of the goniatites was very important to geologists. From the Devonian onwards, until the end of the Mesozoic, these and their descendants are the most useful fossils for dating the rocks. They swam freely in the sea and so were widely distributed; at the same time they evolved rapidly, constantly producing new genera and species. Some of the earliest goniatites can be found at Saltern Cove, just west of Torquay, but they are very small and not impressive to the non-specialist.

There are volcanic lavas and ashes in the Devonian rocks of both north and south Devon, but they are difficult to observe. They include pillow-lavas obviously extruded under the Devonian sea. The volcanic rocks are much better seen in Cornwall, notably near Tintagel and St. Minver. Pentire Head at the mouth of the Camel River, shows a magnificent section in Upper Devonian pillow-lavas. Much of Cornwall is made of highly disturbed sedimentary rocks. They are usually completely lacking in fossils, but it is probable that they are all Devonian in age.

THE CARBONIFEROUS PERIOD

Below the Clifton Suspension Bridge at Bristol is a deep gorge cut by the River Avon. Some distance downstream from the bridge, low bluffs of Old Red Sandstone can be seen beside the river. Between the bluffs and the bridge is what is probably the finest and most complete section in the Lower Carboniferous in Europe. Beyond the bridge is the city of Bristol, built on the Upper Carboniferous with its valuable coal-seams, but the cliffs of the Avon Gorge are all of limestone—the Carboniferous Limestone—packed with fossils (see Plate 6). These limestones were deposited in a clear, shallow sea which spread in from the west over the Old Red Sandstone continent. The fossils are all animals which lived in shallow water and include vast numbers of corals. There are quarries at short intervals along both sides of the river, though those on the left bank are more accessible than those opposite. Studying these, a local amateur geologist, Arthur Vaughan, showed in 1905 that the thick limestone could be subdivided by means of its corals and brachiopods. Vaughan's subdivisions have now been applied all over the country, but they have had to be modified and reinterpreted, for static organisms such as corals and brachiopods are not usually very good horizon makers.

Before Vaughan's time the Carboniferous Limestone was known as the 'Mountain Limestone'. This name comes from the fact that some of the chief upland areas of the British Isles are made of this formation. The Pennines, the Peak District and the Mendips are notable examples.

The Mendips are only about fifteen miles south of Bristol and show the same rocks as the Avon Gorge. They extend as a steep-sided ridge westwards from Frome to the coast, and can be seen heading across the Bristol Channel towards South Wales as the islands of Steep Holme and Flatholm. The Mendips contain many geological wonders. The limestone is easily dissolved by running water and is full of cracks and caverns. Wookey Hole near Wells is a long series of huge underground caverns, made by a stream which still runs through them. The outermost caves were occupied by prehistoric men in geologically recent times. Farther west, like a deep winding wound on the south flank of the hills, is Cheddar Gorge, swarming with summer coaches. Here too, there are stalactite-hung caves, and the gorge, with one wall sheer and the other sloping inwards, may itself be a vast collapsed cavern. There are scores of other, less commercialized caves and pot-holes all over the Mendips, and also the remains of many old lead mines, some of which were worked right back in Roman times.

The Mendip limestone extends across the Bristol Channel and forms an escarpment all round the South Wales coalfield. In Derbyshire, a great dome-shaped structure brings the Carboniferous Limestone up from beneath its cover of later rocks and so forms the Peak District. Here, besides the limestone and some thin shales, there are volcanic rocks in the Carboniferous and the actual vents of volcanoes can be recognized in places. This area too is famous for its caves and pot-holes in the soluble limestone and there are also many mines, some of great age, dug in the veins of lead ore associated with the volcanicity. A mineral for which Derbyshire is particularly famous is fluor-spar or 'Blue John'. This is an attractive semi-transparent mineral, varying in colour through blues, mauves and pinks; it was used for making rather ugly ornaments. Fluor-spar is often found in the lead mines, but formerly most of it was thrown on the tips as valueless. Nowadays the lead mining is no longer worth while, but the tip-heaps are being sorted over for the valuable fluor-spar, which is used for various chemical processes.

Between the Peak District and the Carboniferous areas around Bristol and South Wales, the Carboniferous Limestone thins out and disappears (see Figure 27). Here was the 'Midland Barrier' which separated the seas which invaded the Bristol Channel area from those which spread eastwards over northern England. To the north of Derbyshire in Lower Carboniferous times there was a trough in which accumulated thick muddy sediments. North again, in the Craven Uplands of Yorkshire, there are clear-water limestones as in the Peak District. Apparently the limestones were laid down in shallow seas on either side of a deeper-water trough. Along the margins there were flourishing reefs. These were not coral reefs as we know them today, but mounds on the sea-floor in which almost every type of marine organisms excelled in size and abundance. In the muddy deeper waters, goniatites were common. Unfortunately they are extremely rare in the south-west of England and so are difficult to correlate exactly with Vaughan's coral/brachiopod zones of Avon Gorge.

Similar limestones to those of the Craven Highlands spread into the north-east corner of Wales where they form the magnificent escarpment at Llangollen (see Plate 7). Farther north in England the clear-water limestones split up into thin bands, separated by sandstones and shales. During Lower Carboniferous times, conditions here were constantly changing—sometimes clear seas, sometimes

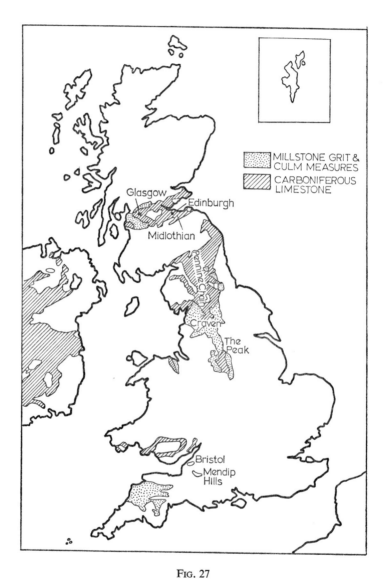

Fig. 27

Map showing the main outcrops of Carboniferous rocks older than the Coal Measures in Britain.

muddy seas and sometimes sandy deltas. Clearly this was near the coastline of a contemporary continent, where the sediments being brought down from the land periodically built up faster than the sea-floor subsided to receive them. In what is now Northumberland, the process went a stage farther and at times the sandy deltas must have emerged from the sea and thick vegetation grew on the exposed swampy, low-lying surface. The dead vegetation formed peat, which is now transformed into coal. The main coalfields of Britain are in Upper Carboniferous rocks, and the subject will be more fully discussed later, but in Northumberland these earlier coals are also important.

The Lower Carboniferous coals continue into eastern Scotland where there are also **oil shales** at this level in the Lothians and Fife. These are shales impregnated with oil which was extracted by distillation. The oil seems to have come from drifted plant debris which accumulated in estuaries and lagoons. In this way the oil shales differ from the coal seams which were formed from plants actually growing more or less in the position where they are found.

At the same time as this sedimentation, there were great outbursts of volcanic activity in southern Scotland. In the west—around and below the city of Glasgow—is a wide sweep of lavas which were piled flow upon flow. They are best seen in the Campsie Fells, north of the city. The individual lava flows often have reddened and decayed tops, which show that they were subjected to the effects of the weather before the next flow came to cover them. In places the actual vents can be seen, for example Dumbarton Rock which is an upstanding pinnacle of solidified lava. In the east there are many such infilled necks, notably in the city of Edinburgh itself. The huge rock from which the ancient castle dominates the Scottish capital has long been important in human history, but its geological history is vastly longer. Scores of millions of years before the first defences were built on its summit, Edinburgh Castle Rock formed the core of a Carboniferous volcano. Similarly, overshadowing Holyrood Palace and dwarfing its few hundred years of recorded history, are the hills of Arthur's Seat. This was a volcano which exploded in Carboniferous times in very much the same way as Vesuvius did when it smothered Pompeii in August, A.D. 79.

All these volcanic extrusions and intrusions are mixed up with the alternating estuarine and shallow sea deposits of the Midland Valley. On the north side of the Firth of Forth there are fresh-water lime-

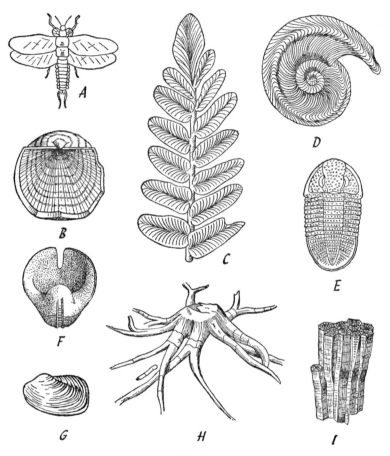

FIG. 28

Carboniferous fossils. *A*: primitive insect (after Handlirsch), $\times \frac{1}{5}$; *B*: brachiopod, $\times \frac{1}{2}$; *C*: compound leaf of a seed-fern, $\times 1$; *D*: early ammonoid or goniatite (after Institute of Geological Sciences), $\times 1\frac{1}{3}$; *E*: one of the last of the trilobites, $\times 1$; *F*: marine snail or gastropod, coiled in one plane, $\times 1$; *G*: non-marine bivalved mollusc or lamellibranch, $\times \frac{1}{2}$; *H*: stump and roots of a Coal Measure tree (after Williamson), $\times 1/25$; *I*: part of a compound coral colony, $\times \frac{1}{2}$.

stones, some of which show evidence of local hot springs. In some of these limestones there are beautifully-preserved plant remains, petrified by the mineral-rich waters. The plants are extinct ferns and clubmosses with their cell structures clearly visible when pieces of the limestone are cut and polished. Not far south of the Scottish border is what is perhaps the most impressive of all the Carboniferous igneous phenomena. This is the great Whin Sill which extends right across northern England and finishes on the Northumberland coast with a series of sheer cliffs, the high wall on which stands Bamburgh Castle and Grace Darling's Farne Islands.

The marine fossils of the Lower Carboniferous are, in general, like those of the Devonian, but it does not require a very long geological training before one can easily distinguish between them. The brachiopods, corals and goniatites of the Carboniferous Limestone include many new types (see Figure 28) some of them of large size, especially in the reefs that were mentioned earlier. 'Sea-lilies', which are the plant-like animals known as crinoids, form a large part of some of the limestone bands and produce an interesting effect when polished. A fine display of such crinoid-packed Carboniferous Limestone can be seen in the Royal Festival Hall in London.

Above the Carboniferous Limestone in England comes a varied series of shales and sandstones, usually called the **Millstone Grit**. Presumably the land around the Carboniferous Limestone sea was elevated so that muds were carried down and deposited instead of the clear limestones. The shales contain marine fossils, notably goniatites, by means of which they have been very finely divided. The same zones which were first recognized by W. S. Bisat in the north of England have since been recognized, with slight modifications, as far away as Siberia and Indo-China.

The sandstones of the Millstone Grit are vast delta and flood plain deposits laid down by rivers flowing off the newly-elevated land areas. The term **'grit'** is used for sandstones in which the grains are sharp and angular. Such sandstones were long ago found to be suitable for making millstones, hence the name for the formation. The hard sandstone beds form steep moorland escarpments with crags and quarries, whilst the shales are hardly exposed at all in the lowland country between, save in the banks of streams.

Studies of the pebbles and sand-grains in the Millstone Grit have shown that most of the material was derived from rocks like those seen in the Scottish Highlands at the present day. It is therefore thought that the sandstones were laid down by a great river or system of rivers coming from the north and north-east. The fact that the false-bedding is usually inclined towards the south-west confirms this idea. The shales, on the other hand, were laid down by incursions of a muddy sea from the west.

COAL MEASURES

UNDERGROUND EXTENSION
OF YORK, DERBY AND
NOTTINGHAM COALFIELD

Fife
Lanark
Edinburgh
Ayr
Northumberland
& Durham
Cumberland
Yorks.
Lancs
Flint
N.Staffs
S.Derby &
Leicester
Shrewsbury
Coalbrookdale
Warwick
S.Staffs
Forest
of Wyre
Pembroke
S.Wales
Forest of Dean
Somerset
Kent

FIG. 29

Map showing the main outcrops of Coal Measures in Britain. It should be noted
that the Kent coalfield is completely concealed beneath younger rocks.

DRY LAND AND SHALLOW SEAS

Walking east or west from the Pennine backbone of England and crossing the successive Millstone Grit escarpments, one comes to the torn countryside of the Coal Measures. A few years ago a prominent politician made a remark that Britain was made of coal and surrounded by fish. We are only concerned here with the first of these attributes, and though Britain is not literally made of coal, there certainly is, or has been, a great deal of it in these islands. The basis of Britain's industrial wealth was the geological accident of a coincidence of coal and iron deposits close together and near the surface in many parts of the country. A glance at Figure 29 is sufficient to show how fortunate we are compared with a country such as Italy which has practically no coal deposits at all. Human history would have been very different had the use of underground oil been discovered and applied before that of underground coal.

The **Coal Measures**—the thick series of strata that contain the majority of our usable coal—come at the top of the Carboniferous System in Britain. In other parts of the world, the chief coal seams occur in later strata, but a great belt of Upper Carboniferous Coal Measures extends right across the northern hemisphere from the mid-west U.S.A. to the Donetz basin in southern Russia. Actual coal forms only a tiny proportion of this formation, the greater part is sandstone and shale.

The Coal Measures were laid down on a surface which oscillated just above and just below sea-level. Many **cycles of sedimentation** are recognized in these deposits. Each cycle, when fully developed, started with an invasion of the sea over the low-lying land when a small thickness of marine sediments was laid down; then the deposits built up above sea-level and the sediments with marine fossils were followed by river and swamp muds, often packed with the remains of land plants and fresh-water molluscs. Above these came river and deltaic sands and then a coal-seam. The cycle was then completed by a return to marine conditions.

The bed immediately below a coal-seam is often what is called a **seat-earth**. This is the actual soil in which the coal plants once grew. Seat-earths are characteristically deficient in the mineral salts which a plant removes from the soil and they also sometimes contain the stumps and roots of trees still in the position in which they grew. The Coal Measure plants apparently flourished in dense swampy forests with some of the plants partly submerged in water. One can trace the ephemeral streams which wandered about through the

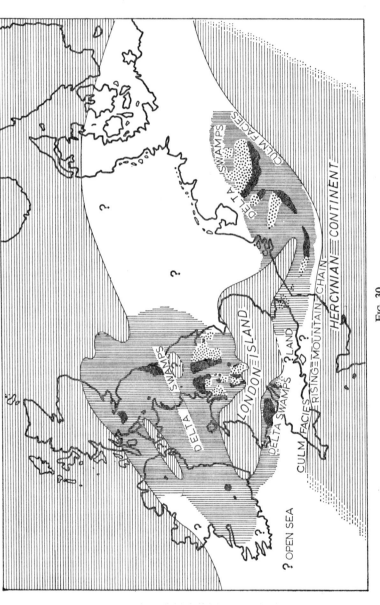

Fig. 30

Map showing the supposed geography of north-west Europe during Coal Measure times. Exposed coalfields are shown black, and concealed coalfields stippled. (Slightly simplified from *A Palaeogeographical Atlas of the British Isles . . .* by

swamps. These are preserved as what miners call 'wash-outs', where the coal-seam is suddenly cut out by a sandy deposit filling a channel (see Figure 7 top). As the plants died, they fell and formed a layer of partly decayed vegetable matter or peat, which slowly thickened as generation followed generation. Comparable deposits are seen today over large areas of Scotland, Ireland and upland England, though these are comparatively lacking in woody material. With the passage of time and the burial of the peat beneath thick later sediments, it was greatly compressed and lost much of its gaseous content. In this way it was transformed first to brown coal or lignite and then to the black coal which many of us still burn so wastefully in our fireplaces. In certain areas, the process goes a stage farther, either because of a heavier load of sediments or the proximity of earth-movements. Then the ordinary bituminous coal is transformed into shiny black **anthracite** which does not soil the hands when touched. Anthracite has a very low gas content and is a very valuable deposit.

The thickness of a coal-seam depends largely on how long plants could grow on the swampy surface before it was submerged again and covered with other sediments. Theoretically each cycle ends in this way with a marine transgression, but few of the Coal Measure cycles are complete, and the marine bands in particular (with their very useful fossils) are often wanting. In the upper part of the Coal Measures there are none at all, implying that the sea had been completely excluded from the British area.

The fossils in the non-marine parts of the Coal Measures are mostly 'mussels' and plants. The 'mussels' are bivalve molluscs or lamellibranchs quite like animals which live in modern rivers and lakes. Their shells are rather featureless, and are usually found flattened in the dark shales, but even this unpromising material can be used stratigraphically. The plants are, of course, the basis and the chief interest of the Coal Measures. In the coal itself, not much can be seen of the plant structures, though if it is broken down with various chemicals it is often found to contain millions of minute spores, and these can be used for correlating the seams. In the shales there occur pieces of bark, leaves and other parts of a wide variety of plants. Most of them can be shown by their detailed anatomy to be closely related to forms such as the 'horse-tails' and 'club-mosses' which are living today, but whereas the latter are humble plants never more than a few inches high, their Carboniferous relations were quite large forest trees. Other plants which abound in the Coal

Measures are distant relations of the modern conifers and a large series of forms with fern-like leaves which reproduced by means of seeds (the so-called 'seed-ferns'). The plant remains can be used for sub-dividing the Coal Measures though not so finely as with other fossils.

The British coalfields fall naturally into four main regions: the southern coalfields, the Midland coalfields, the Pennine coalfields of northern England and those of the Midland Valley of Scotland. These are all shown in Figure 29.

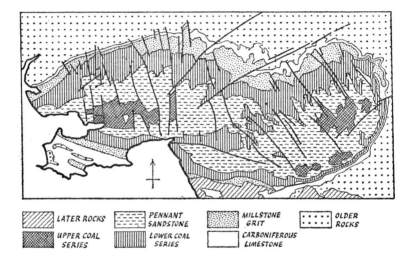

LATER ROCKS PENNANT SANDSTONE MILLSTONE GRIT OLDER ROCKS

UPPER COAL SERIES LOWER COAL SERIES CARBONIFEROUS LIMESTONE

FIG. 31

Geological sketch-map of the South Wales coalfield (re-drawn, slightly simplified from British Regional Geology Handbook *South Wales* by permission of the Controller of H.M. Stationery Office).

In the south there are the coalfields of Pembrokeshire, South Wales, the Forest of Dean, Somerset and Kent. Of these, the South Wales coalfield is by far the most important. It forms a great oval basin, longest in an east/west direction (see Figure 31). In the middle of the Coal Measures here there is a thick sandy formation known as the Pennant Sandstone. This dominates the local scenery, since it is far more resistant to weathering than the shaly formations above and below. The main coal-seams are below the sandstone and so the coal mines are mostly placed in narrow steep-sided valleys cut down through the Pennant formation. The famous crowded mining valleys

112

such as the Rhondda and Ebbw Vale are separated by open, upland moors. Besides the coal, there are a great many thin bands of iron-stone in the Welsh Coal Measures. These are no longer of any value, but they were formerly worked on a big scale, even before the days of coal-mining, and centres such as Merthyr Tydfil owe their origins to the local presence of iron.

The much smaller coalfield of Pembrokeshire differs from the main coalfield in that it has been more severely disturbed by later earth-movements. It also differs in that its marine bands are much better developed, suggesting that the short-lived invasions of the sea in Coal Measure times came from this direction. The Forest of Dean in Gloucestershire only contains the upper part of the Coal Measures, preserved in an asymmetrical syncline. Many of the coal seams here are thin and have never been worked on a large scale but the whole region is pitted with the small workings of the 'Free Miners' who have ancient rights dating back at least to the thirteenth century. The Somerset coalfield consists of several basins, one of them extend-ing right under the city of Bristol, but most of the workable coal has been removed. The succession is very like that of South Wales except that there is a much thicker development of the measures above the Pennant Sandstone. The scenery, however, is quite different; instead of the deep valleys and upland moors, there is a gently rolling low-land countryside. This has nothing to do with the rocks themselves, but results from later geological history. South Wales has been raised several hundred feet within the recent geological past and has been deeply dissected by weathering, whereas the Somerset area has hardly been elevated at all.

Eastwards from here practically nothing is known of the Coal Measures until we reach Kent. Here we have the only British example of a completely concealed coalfield, that is, a coalfield of which there is no evidence at all at the surface. The Coal Measures in Kent are buried beneath about 1,000 feet of much younger sediments. Their presence was forecast, on the basis of good geological reasoning, by Godwin-Austen in 1856, but they were not proved until a boring was put down at Dover in 1890. The succession of rocks is again very much like that in South Wales, and we can presume that there was once a more or less continuous belt of Coal Measures being deposited right across from west Wales to east Kent and then on into the continent (see Figure 30).

In the English Midlands the Coal Measures dominate the geology, the scenery and the life of the people. The Midland coalfields radiate like the fingers of a hand from the south end of the Pennines. If the North Staffordshire coalfield is the palm of the hand, then the disused Shropshire coalfields are the thumb, the Coalbrookdale and Forest of Wyre coalfields are the first finger, the middle finger is the important South Staffordshire field, the third finger is the Warwickshire field and the little finger is the small coalfield shared by Leicestershire and South Derbyshire. With the coalfields of the Midlands we can conveniently group the coalfields of north-east Wales which form a long belt running south from the Dee estuary. All these separate areas were probably contained in a single basin of deposition in Upper Carboniferous times. Deep borings have shown that in many places the Coal Measures continue between the coalfields, but at a great depth below later deposits.

Apart from the coal-mining itself, many of Britain's major industries are associated with the Midland coalfields. The North Staffordshire field, for example, is also known as the 'Potteries Coalfield', for here are situated Arnold Bennett's 'Five Towns'—Stoke-on-Trent and the rest—in a hollow filled with the smoke of the great china factories. The most important of the Midland coalfields in the past was undoubtedly that of South Staffordshire, though it is now very much on the decline. Here the proximity of coal and ironstone in the Coal Measures, together with other geological factors, produced the iron manufactories of the Black Country, and that city of innumerable industries—Birmingham.

The third area of Coal Measures in England comprises the four coalfields which are arranged in a regular pattern on either side of the Pennines. To the west and east respectively at the southern end of the chain are the Lancashire-Cheshire coalfield and the huge Yorkshire-Derby-Nottingham coalfield. These are the most important of our coalfields today and support some of the largest industrial areas. The latter is particularly important, especially as it extends eastwards under younger deposits with vast untouched reserves (see Figure 29). Many beds, including individual coal-seams, can be correlated across the Pennines, showing that the Coal Measures once extended right over this area. Immense thicknesses of sediment must have been removed by erosion when the Pennines were later uplifted. These two coalfields have much in common with those of the Midlands and were probably laid down in the same basin.

The coalfields at the north end of the Pennines are more like those of Scotland. These are the Northumberland and Durham coalfield to the east and the Cumberland coalfield to the west. The former is an almost complete structural basin, but the latter dips westwards below the Irish Sea and coal-seams are worked far out under the sea-floor. The Northumberland and Durham coalfield was formerly of great importance to London because of the ease with which its coal could be carried there by sea down the east coast. This 'sea-cole' from Tyneside was unpopular for domestic use until wood became scarce in Tudor times; it was heavily taxed for various purposes including the rebuilding of St. Paul's after the great fire of London.

In Scotland, the Upper Carboniferous Coal Measures extend across the Southern Uplands via the small Sanquhar and Thornhill basins and occur in scattered patches all along the Midland Valley. Here again they had much to do with the industrialization of the area, but it should be remembered that in Scotland and Northumberland there is also valuable coal in the Lower Carboniferous rocks.

A feature of the Upper Carboniferous which becomes increasingly apparent the farther one goes north, is the presence of what are generally called **red beds**. These are red shales and sandstones, lacking in coal, which as a generalization may be said to have appeared early in the north, but progressively later farther south, so that in South Wales there is only a trace of them at the very top of the succession. In places these red formations contain beds which seem to have dried up and cracked under a hot sun, and the general red colour of the sediments is thought to indicate the oxidizing conditions of an arid climate. These features foreshadowed what was to follow in the next period and this will be described in Chapter VI.

One should not write about the geology of Britain today without saying something about the vast resources of oil and gas that have been found in the North Sea during the last ten years. These may well change the whole economic future of Britain just as did the development of coal-mining at the beginning of the last century. It is interesting to note therefore that the Carboniferous rocks are the source of most of the oil and gas as well as most of the coal. But oil and gas are much more mobile than coal and they tend to move upwards and accumulate in the more porous rocks, most notably in the red sandstones of the Permian and Triassic which will be discussed in the next chapter.

VI

MOUNTAINS AND DESERTS

This chapter deals with the Hercynian Orogeny which reached its climax at the end of the Carboniferous Period, and with the Permian and Triassic Periods which followed. It therefore covers the change-over from the Palaeozoic to the Mesozoic Era.

THE HERCYNIAN OROGENY

The Coal Measure swamps came to an end with the second Palaeozoic paroxysm of mountain-building. These movements were the Hercynian Orogeny, which takes its name from the Harz mountains in Germany, where its effects are particularly clear. It is also commonly known as the Armorican Orogeny after Armorica or Brittany.

The first shudders of the mountain-building movements occurred long before the end of the Carboniferous—perhaps as far back as the Devonian. There are breaks here and there within the Carboniferous succession which reveal such minor tremors. Thus near Malham in Yorkshire, a fault can be shown to have affected the Carboniferous Limestone, but the Millstone Grit passes straight across it without any signs of disturbance. In other places—notably along the line of the Malverns—there was uplift, erosion and even folding before and within the Coal Measures. But such disturbances were merely curtain-raisers for the violent convulsions which came at the end of the Carboniferous.

The Hercynian Orogeny produced its most violent effects and is best studied on the continent. Whereas the Caledonian folding had been mainly on north-east/south-west lines, the Hercynian folds were dominantly east/west (see Figure 32). The only part of Britain where these can be seen clearly is the south-west. The Devonian and

GRANITE

Bodmin Moor
St. Austell
Land's End
Scilly Isles
Dartmoor

FIG. 32

Map showing the main trends of Hercynian folding in Britain, with some of the granite masses emplaced at this time.

Carboniferous rocks of Devon and Cornwall were very strongly folded at this time (see for example Plate 9). The muddy sediments have been altered into slates, as in the older rocks of North Wales, and the fossils are often deformed and obliterated. Thus at Delabole in North Cornwall, muddy Devonian sediments have been converted into slates and are still worked in a deep and dangerous quarry. A common brachiopod in these beds is called the 'Delabole butterfly' by the quarrymen because of its flattened appearance.

The front edge of the Hercynian mountains was probably advancing on Britain from the south during Upper Carboniferous times (see Figure 30). At the climax of the orogeny it must have run roughly along the line of the Bristol Channel and on across the south of Ireland. North of this line the effects of the orogeny are far more subdued and consisted mainly of up and down movements and faulting (see Plate 8). Along the actual mountain front there was considerable overthrusting, comparable to the Caledonian thrusts of north-west Scotland. Thus at Vobster quarry near Radstock in Somerset, the Carboniferous Limestone has been pushed right over the Coal Measures. In Belgium the same thing has happened on a bigger scale; there are thrusts between the Dinant and Namur coalfields, and the Campine coalfield in the north is completely hidden under overthrust masses.

The east/west trend of the Hercynian folds shows up clearly on a geological map of southern Britain (see for example the South Wales coalfield—Figure 31). It is also revealed by the deeply dissected coastline of Kerry and Cork in south-west Ireland. Here the harder Old Red Sandstone in the cores of the anticlines forms the westerly-projecting peninsulas, whilst the softer, synclinal Lower Carboniferous rocks form the deep inlets. In south Pembrokeshire and the Gower Peninsula near Swansea, the Carboniferous rocks were closer to the mountain-front and so were more affected than other parts of Wales. Again the Old Red Sandstone is more resistant to erosion (see Plate 17). Farther north in Britain the trend of the folds swings from east/west to north/south. This may be the effect of the rigid mass of older rocks in Wales. Thus in Midland and northern England we have the structures of the Malverns, the coalfields and the Pennines which are mainly at right angles to those of the south-western peninsula (Figure 32).

Some geologists see the effects of the Hercynian Orogeny as resulting from the closing of an east-west late Palaeozoic ocean across the centre of what is now Europe, with subduction zones going down under south-west England on one side and Brittany on the other. Certainly we have some pillow lavas of late Palaeozoic age, for example at Chudleigh in Devon and at Pentire Head in Cornwall, but others would argue that there is no real evidence of oceanic crust across Central Europe in Devonian and Carboniferous times and the subduction zone was way down south in the Mediterranean area.

In the Hercynian Orogeny, as in the Caledonian, the most intensive deformation of the rocks was accompanied by the emplacement of granite masses. These are well seen in Devon and Cornwall, where there are a series of granites exposed today after long ages of erosion. They stand up as upland moorland areas, characteristically surmounted by blocky 'tors' (see Plate 12). The largest of the granite masses is Dartmoor, which covers an area of about 240 square miles. Smaller masses to the south-west are the Bodmin Moor, St. Austell, Falmouth, Land's End and Scilly Isles granites, and there are several smaller ones such as St. Michael's Mount near Penzance. The granites are arranged along a north-east/south-west line and are uniform in composition. They are obviously closely connected with one another and it may be that there is, in fact, just one huge granite which has only been partly unroofed.

After the granites had been emplaced, they were altered by the action of hot vapours rising from below. These introduced new minerals such as tourmaline and topaz. Most important was the alteration of the felspar (which forms a major part of the granite) into kaolin or 'china clay'. This happened particularly in the St. Austell granite. The china clay is worked on a huge scale for use in the best types of pottery.

Also running obliquely through Devon and Cornwall is a belt of mineralization about ten miles wide which can be blamed on the Hercynian Orogeny. In this belt there occur many ore-bodies, filling fissures or replacing the original rocks, and yielding many different valuable minerals. The tin, copper and lead of Cornwall are the most famous, but there are many other metalliferous ores, including even a little uranium. Tin is also found as pebbles in stream deposits, derived from the weathering of ore bodies, and it has been recovered from stream gravels since prehistoric times. Cornish tin was in fact responsible for Britain's first mention in recorded history, when Phoenicians (or even earlier peoples) came for it to a trading post probably situated on St. Michael's Mount. The mining industry in Cornwall today is but a shadow of what it was, due to the discovery of more easily worked deposits in other parts of the world, notably the tin of Malaya.

Apart from these violent but localized effects of the Hercynian Orogeny on Britain, the most widespread effect seems to have been one of general uplift. This has already been mentioned in connection

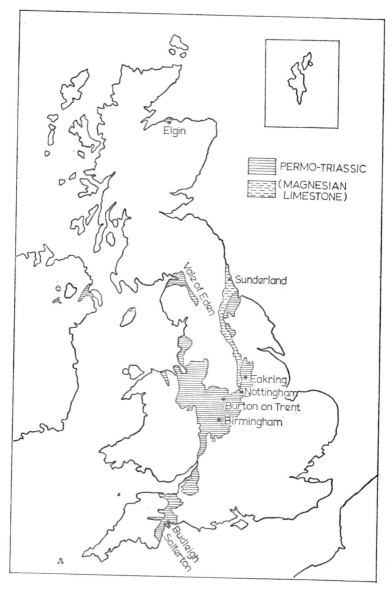

FIG. 33
Map showing the main outcrops of Permian and Triassic rocks in Britain.

with the Carboniferous succession. It culminated in the complete exclusion of the sea and in the semi-arid 'red-bed' conditions of the uppermost Coal Measures. In many areas it is extremely difficult to separate these from the Permian and Triassic rocks which followed.

The newly-raised mountains to the south of Britain appear to have formed a barrier against the rain-bearing winds coming from the ocean to the south. Thus the whole of northern Europe in Permian and Triassic times may have been in a rain-shadow area like that of the west-central states of the U.S.A. today. In this arid climate, the sediments which accumulated were characteristically red in colour. This is commonly attributed to the rapid oxidation of iron minerals under such conditions, producing iron oxides comparable to the red rust on neglected tools. The process was probably more complicated than this, but there are many other lines of evidence indicating an arid desert climate at this time.

The red colouration is the most obvious feature of the Permian and Triassic rocks of north-west Europe and they are commonly spoken of together as the 'New Red Sandstone', in contradistinction from the 'Old Red Sandstone' which preceded the Carboniferous. All these rocks are notoriously lacking in fossils and are subject to considerable lateral change, so that it is extremely difficult to correlate them from place to place, or even to draw a line between the Permian and Triassic. This is particularly unfortunate since these two systems mark the end of the Palaeozoic and the beginning of the Mesozoic respectively. This is therefore one of the points in the stratigraphical record where we cannot afford to be insular.

In Britain the Permian and Triassic rocks are essentially a great thickness of red, unfossiliferous sediments coming between the Coal Measures and the marine deposits of the Jurassic. They cover very large areas and their study is important economically, firstly because of what they contain themselves and secondly because their thickness is an important consideration in connection with the mining of the coal which they often cover.

THE PERMIAN PERIOD

In 1841 the Czar of Russia invited the most eminent geologists of Britain, France and Germany to come and study the then almost unknown geology of his country. Sir Roderick Murchison had then completed his great researches on the Palaeozoic rocks of this country and was the obvious man for the task. When he went to Russia, Murchison found a tremendous thickness of varied rocks coming on top of the Coal Measures, and he called these the Permian after the province and city of Perm. This name soon passed into the geological literature.[1]

Much of what Murchison called Permian would now be grouped into the Carboniferous, but Russia provides a useful standard for the system, since the succession there contains a much more complete marine record than is known in north-west Europe. Around Moscow, the Carboniferous and Permian chiefly take the form of shallow water limestones, but farther east, in the Urals, they pass into a very much thicker geosynclinal facies. The Urals were a huge north/south geosynclinal trough, connecting with an open ocean to the south and receiving thousands of feet of sediment during late Palaeozoic times.

In the Permian strata of Russia there are huge marine faunas of Palaeozoic type. The most useful fossils are spindle-shaped unicellular foraminifera known as **fusulinids** (see Figure 35). These are closely related to Carboniferous types, as are other fossils such as brachiopods, corals and echinoderms. There are also goniatites and the first of a related group of cephalopods—the **ammonites**—in which the partitions between the chambers are folded in a very complex fashion. Similar fossils are found in the thick marine Permian rocks in the south of Europe, and a deep broad sea is postulated in that area. This sea is known by geologists as the **Tethys** and will figure in much of the later part of this history. The Tethys extended in an east-west direction around the northern hemisphere. The modern Mediterranean may be regarded as its only surviving vestige, but from late Palaeozoic until early Tertiary times it formed one of the main elements in world geography. In Permian times it stretched eastwards through the Middle East, northern India and on to southeast Asia and the East Indies. The island of Timor contains one of the best developments of the system with beautifully preserved faunas of specialized Palaeozoic types. In the opposite direction, the Tethys extended into southern U.S.A. and an almost enclosed basin, presumably opening off it, was developed in Texas and adjoining states. Here thousands of feet of sediment accumulated and have

[1] Perm was later renamed Molotov, which, since it means 'man of the hammer' might make an even better name for a geological system. However, political vicissitudes have brought about a reversion to the old name once more.

been studied in great detail because they form a valuable oil reservoir.

In north-west Europe, however, the situation is not so clear. It is difficult to separate the Permian from the Carboniferous and even more difficult to separate it from the Trias. In most parts of Britain the Permian consists of breccias and conglomerates interbedded with red desert sandstones. These are the products of the rapid mechanical weathering of the newly-raised Hercynian mountains. In the Vale of Eden, for example, between the Lake District and the Pennines, a very close parallel can be drawn between the local Permian deposits and the conditions in certain inland basins in Central Asia at the present day. The latter have an arid climate and are bordered by steep mountains from which great fans of scree spread out on to the desert sands of the valley floor.

In the Midlands, the Hercynian movements continued, chiefly as repeated movements along the faults which bound the various coalfields. Apparently the coalfields stood up as rigid masses whilst the areas between were down-faulted troughs. Thus the troughs received greater thicknesses of Permo-Triassic continental deposits and the Coal Measures between the exposed coalfields are buried at great depths. When faulting occurred, the upstanding masses suffered very rapid erosion and thick conglomerates and breccias were deposited very rapidly. The former seem to have been formed by torrential rushes of water like the seasonal downpours which periodically sweep into modern rain-shadow areas. It can be shown in places that the older breccias and conglomerates contain fragments of Silurian and Carboniferous rocks, whilst the younger debris comes from Precambrian and Cambrian sources. This illustrates the stripping away of the younger rocks and the uncovering of the older cores of the mountains.

The Permian rocks in north-west Europe were not entirely of continental origin; a sea did for a time extend over the area in late Permian times. This was an unusual, enclosed sea and probably had only a restricted connection with an open sea—usually called the **Boreal Ocean**—somewhere to the north of Europe. Its fauna, though abundant in places, was of very limited variety and many of the characteristic Tethyan creatures were absent. This sea extended across Germany and into north-east England, where there was laid down a deposit known as the Magnesian Limestone (see Figure 34). Normal limestone consists almost entirely of calcium carbonate, but this Permian limestone contains a high proportion of magnesium carbonate. In places there are peculiar magnesium-rich concretions, formed after deposition and often resembling fossils. The Magnesian Limestone is quarried on a large scale and has been much used in the past for building purposes. Its most famous use was for rebuilding the Houses of Parliament during the last century. Unfortunately, the geologists' advice was not followed and the wrong beds of stone were used. This had the intriguing result that, due to London's acidic atmosphere, the magnesium carbonate was changed to magnesium sulphate (or Epsom salts) and was rapidly washed away. The amount of repair work that this has entailed was obvious to a generation of Londoners.

The Magnesian Limestone outcrops in a broad belt from Sunderland and the north-east coast (see Plate 10), as far south as Nottingham, and its basal beds rest with very obvious unconformity upon the older strata. For a brief time only, the sea broke through the Hercynian feature of the Pennines into north-west England and thence as far as Ireland. The deposit to the west of the Pennines is insignificant compared with that to the east. Soon after the breakthrough, the sea began to dry up and its margins retreated. With rapid evaporation under an arid climate, the salinity of the sea increased. This had a marked effect on the animals living in the sea; they could not grow properly and the species died out one by one. The increase in the number of 'dwarfs' and the decrease in the total number of species can be demonstrated as one goes higher and higher in the Magnesian Limestone. Eventually life disappeared altogether and the sea became a salt lake like the Dead Sea.

In this lake, rock-salt was deposited together with a whole series of other minerals known as **evaporites**. These are the various salts which can be obtained by the evaporation of sea water, and some of them are very valuable. Rock-salt and deposits such as gypsum and anhydrite have been known and worked in the British Permian since Roman times, but rarer salts of magnesium and potassium have only been discovered since the last war, in deep borings in east Yorkshire. The same salts are known and mined on a large scale at Stassfurt in Germany. It is probably significant that the rarer salts, which require the evaporation of fantastic quantities of sea water, are only found over a more restricted area than is common rock-salt. They may represent the final drying-up of the Permian sea. These salts are extremely soluble in water and are only preserved when they are covered with an impervious layer such as clay.

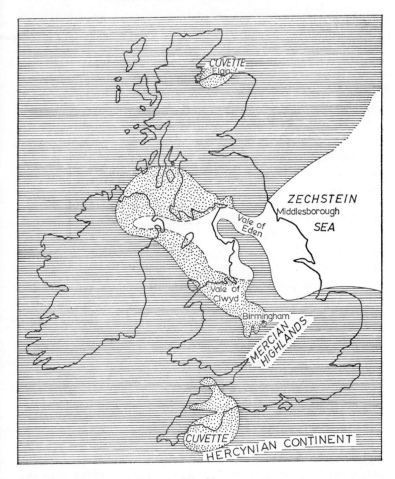

FIG. 34

Reconstruction of the supposed geography of late Permian times. (Simplified from *A Palaeogeographical Atlas of the British Isles* . . . by L. J. Wills, by kind permission of the author and Messrs. Blackie & Son, Ltd.). The stipple indicates areas of land deposition.

Red beds with salt deposits come on top of the thick marine successions in both the Urals and the Texas region, so that even in these classic areas the story is not complete and it is difficult to be sure where the Permian ends and the Triassic begins. The Tethys also retreated at this time and evaporites spread as far south as Austria,

125

where Permian salt has long been a valuable commodity. At Halstatt in the Tyrol, salt has been mined since before the coming of the Romans. Old adits have been found filled with the burnt spills held by prehistoric miners in their teeth while they dug the salt with primitive tools. The name of Halstatt (or 'salt town') is well known to archaeologists for the remarkable series of early Iron Age cultures left by the settled community of salt miners.

By the end of Permian times in Britain, much of the country was covered by a hot sandy desert. West of Birmingham there are extensive outcrops of a formation known as the 'Lower Mottled Sandstone'. This is variously regarded as late Permian and early Triassic in age. The formation has been studied in great detail and proved to be very like the sand found in modern deserts. It is often strongly false-bedded and this is not the deltaic type of false-bedding produced by water deposition, but the false-bedding of desert sand-dunes (as in the Saharan dune seen in Plate 20); similar dune-bedding in a Scottish Permian sandstone is seen in Plate 13. Each sand-dune has a gently-sloping surface on the windward side and a steeply-sloping surface away from the wind. Thence, from a large number of observations, it has been possible to determine the direction of the prevailing wind. In the English Midlands, the prevailing wind at the end of Permian times was from the east. Other observations, in Cumberland and in south Devon, show prevailing winds north of east and south of east respectively. The resultant wind pattern is comparable with that of the modern Egyptian desert (see Figure 36).

It has already been shown that the Permian must be placed in the Palaeozoic because of its invertebrate fossils, which are of Palaeozoic type. Some mention should also be made of plants and vertebrates. Not many of the former are known in north-west Europe, presumably because of the unfavourable climate at the time, but those that have been found are closely related to the better known Carboniferous floras. In the Vale of Eden, there is a thin deposit which has yielded plants with thick cuticles and other features suggesting adaptation to a desert environment.

The vertebrates of the Permian are particularly interesting though there are not many known in this country. The best locality in Britain is at Cuttie's Hillock near Elgin, on the south side of the Moray Firth. The surrounding countryside is of Old Red Sandstone and it was long assumed that the sandstones along the coast near Elgin were also Old Red Sandstone. Later, however, bones were found and these proved to be, not Devonian fishes, but Permian and Triassic reptiles.

The first large land animals were the **amphibians** which evolved from lung-fish in late Devonian or early Carboniferous times. Large ungainly amphibians, which could hardly raise their bellies off the ground, are found from time to time in the Coal Measure swamp deposits. They could move, more or less, on dry land, but probably never went far from water, and certainly had to return to it to lay their shell-less eggs. Amphibians were the dominant land animals of the late Palaeozoic, and reached their maximum abundance and variety in the Carboniferous, but by Permian times there had also evolved the next higher grade of vertebrates—the reptiles. These were altogether better adapted for life on dry land and their evolution may have resulted from the expansion of the land areas at this time, just as the land plants evolved when the Old Red Sandstone continent rose above the sea. The early reptiles had limbs capable of raising their bodies above the ground and they could therefore move quite rapidly. They did not have to lay their eggs in water, but the young usually appeared fully-developed from eggs with hard shells laid on the ground.

It is difficult to get a clear picture of Permian land life from the evidence available in north-west Europe, but in other parts of the world, notably in South Africa, Russia and the U.S.A., large land faunas are known. These show that in Permian times the amphibians were still in a dominant position, but they were much reduced in numbers before the end of the period. At the same time there were many different types of reptiles, and some of these flourished in Britain probably because the very dry conditions favoured them but not the water-bound amphibians. With the coming of the Mesozoic, the reptiles were to take the lead everywhere and the amphibians dwindled rapidly to the insignificant position which they hold in the world today.

THE TRIASSIC PERIOD

The second era of life on earth—the Mesozoic—is taken as beginning with the Triassic Period or Trias, though a geologist working in Britain might be excused for wondering why. This major boundary between the Palaeozoic and the Mesozoic is here completely obscured by the after-effects of the Hercynian Orogeny, and even in classic areas such as the Urals and Texas the junction is uncertain. In very few places are marine Triassic rocks with useful fossils found resting on marine Permian, and at least one geologist has suggested that there is no justification for recognizing two separate systems. However, taking the world evidence as a whole, it is obvious that there are Triassic rocks with land and sea faunas very different from those of the Permian.

In Britain the Trias is represented by the upper part of the New Red Sandstone—sandy, pebbly and marly deposits almost completely lacking in fossils. At or near the base there are thick pebble beds in various places. They are best known near Birmingham (where they are called the 'Bunter Pebble Beds') and in south Devon (where they are called the 'Budleigh Salterton Pebble Beds'—see Plate 11). In the absence of contemporary fossils it is impossible to prove that these two deposits are of the same age, but they resemble each other in the nature of the pebbles, which are often several inches across. The coarseness of the deposits suggests that there were further earth-movements at this time—the final tremors of the Hercynian Orogeny—which caused renewed uplift of the mountains and so renewed rapid erosion. The most characteristic pebbles in these deposits are made of a distinctive purplish-coloured quartzite. Patient geologists spend hours breaking open hundreds of these pebbles and eventually, with luck, they may find some rather obscure fossils. Such fossils have proved to be of Ordovician age and their presence is rather a mystery. In the case of the deposits near Birmingham, there are no known local Ordovician rocks which could have been eroded in Triassic times to provide the pebbles. Curiously enough, the quartzite and the fossils match very closely certain Ordovician deposits in Brittany. In some reconstructions of the geography of the time, a great river is shown flowing up from Brittany to Birmingham, but it seems unlikely that large boulders could have been carried so far. An alternative suggestion is that there were similar deposits in southern Britain which are now deeply buried or have been completely eroded away.

Above the early Triassic pebbly deposits in Britain, the sediments tend to become progressively finer grained, through sandstones to dust-like marls. This is, of course, a generalization, but the implication is that the mountains were being worn down, until at the end of the period much of northern Europe was a flat desert plain.

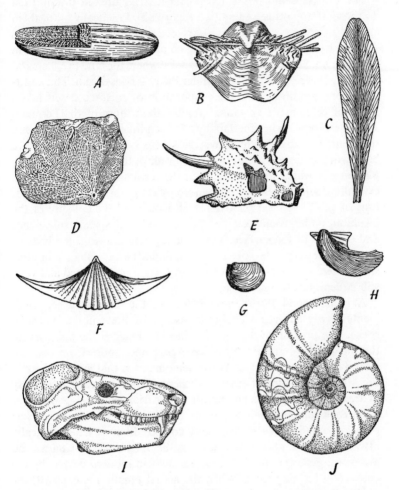

FIG. 35

Permian and Triassic fossils. *A–E*: Permian fossils. *A*: fusulinid, a giant foraminifer (after Zittel), $\times 3\frac{1}{2}$; *B*: spiny brachiopod belonging to a group near
extinction (after King), $\times \frac{2}{3}$; *C*: leaf of *Glossopteris* from the southern hemisphere, $\times \frac{1}{2}$; *D*: polyzoan, a plant-like colonial animal (after King), $\times \frac{2}{3}$; *E*:
skull of an early reptile from Elgin, $\times \frac{1}{8}$. *F–J*: Triassic fossils. *F*: aberrant
brachiopod from the Alps, $\times 2$; *G*: bivalved crustacean from a desert pool,
$\times 2$; *H*: small bivalve mollusc from the Rhaetian beds of England, $\times \frac{3}{4}$; *I*: skull
of a mammal-like reptile from South Africa (after Seeley and others), $\times \frac{1}{10}$
J: ammonite from the Muschelkalk showing characteristic partitions, $\times \frac{1}{2}$.

The Triassic sediments outcrop over a larger area in Britain than do those of any other system (see Figure 33). Practically all of them are red in colour and nearly all are demonstrably of desert origin. The sandstones show dune-bedding comparable to that described in the Permian. Some beds show sun-cracks, where the mud in ephemeral desert pools dried and cracked under the hot Triassic sun. Thick salt-beds were laid down with the evaporation of enclosed desert lakes. Some of these are even more valuable than those of the Permian. The Cheshire plain is underlain by a tremendous thickness of Triassic sediments and these include two thick beds of rock-salt which have been worked since Roman times. Nowadays the salt is obtained by pumping in water and pumping out the resultant brine, from which crystalline salt is obtained once more by evaporation. Thousands of tons of salt have been removed in this way and the collapse of the resultant underground cavities has caused widespread subsidence and damage at the surface. Even where there are no thick beds of salt or other evaporites, one can often find little cube-shaped impressions where isolated crystals of rock-salt were once formed and were subsequently dissolved away.

At Charnwood Forest near Leicester, the Triassic desert sediments slowly covered a craggy landscape of Precambrian volcanic rocks. The present level of erosion has just reached the point where the old mountain-tops are just emerging once more from beneath their Trias cover. Professor Watts demonstrated that we have here in the Midlands an almost exact equivalent of the landscape seen today in areas such as Nevada and South-west Africa, where jagged hills of older rock stand up suddenly from flat desert plains. This is the sort of background which makes many cowboy films at least scenically attractive. Besides the general setting, the older rocks of Charnwood are worn, grooved and fluted as are rocks in modern deserts by the constantly blowing sand. When the actual junction is exposed, the older rocks can often be seen to be highly polished by the sand-blasting, giving them the appearance of what is known as 'desert varnish'. Here and there in Charnwood there are steep-sided valleys cut down into the Precambrian rocks and filled with coarse sediments. These have been compared with the 'wadies' of the North African deserts, which are cut by seasonal torrents, but are usually dry.

Towards the end of the Triassic period in Britain, the mountains were almost everywhere worn down to a flat desert plain, on which

FIG. 36
Prevailing wind directions in the Permo-Triassic desert (redrawn from a map published by Professor F. W. Shotton by kind permission of the author).

there accumulated great thicknesses of red, fine-grained, wind-blown dust. This is known as the 'Keuper Marl' and is very widespread in central and western England.

All the Triassic rocks in Britain are highly permeable and their chief value to man lies in the amount of water they contain. They are second only to the Chalk of south-east England in importance as a water reservoir. This subterranean Triassic water, plus the surface water from rivers and lakes so much used in the Midlands, is preferable to the Chalk water which has to be endured by Londoners. It has also proved particularly suitable in certain areas for brewing and other processes which need a large water supply. The famous breweries of Burton-on-Trent are so situated because of the local Triassic water supply, and it is said that certain salts present in the rocks give a distinctive and desirable flavour to the beer. All facets of human life are closely controlled by geological phenomena.

Apart from plant spores almost the only fossil one can hope to find in these rocks is a little bivalved crustacean which occurs crowded together in large numbers. These creatures are very like a form which lives today in desert pools; it presumably lived in the same way in the pools which came and went in the Triassic desert. Much rarer are the remains of Triassic plants, fish and land vertebrates. The Permian vertebrates of Elgin have already been mentioned. At nearby Lossiemouth (chiefly remembered as the home of Ramsay Macdonald) similar sandstones have yielded Middle Triassic reptiles. In other parts of the world, notably in South Africa, larger faunas of Triassic reptiles are known and this gives us a clear picture of the stage reached in vertebrate evolution at that moment in time.

There were still a fair number of large, ungainly amphibians, but these were highly specialized forms, soon doomed to extinction. Reptiles were by now very much in a dominant position, and included the ancestors of many new major groups. There were reptiles that had adopted the habit of walking on their hind-legs and so leaving their fore-limbs free for feeding purposes. These were probably ancestral to the dinosaurs which were to be so important in later Mesozoic times. Other reptiles, after becoming well adapted to life on dry land, had reverted to an aquatic life. Most important of all was a group of small reptiles probably ancestral to the mammals. Some of them have in fact been called mammals, but it is probably wiser to refer to them as mammal-like reptiles. The essentially mammalian features of warm blood, hair, live birth, suckling young and so on, are not such as would normally fossilize. The palaeontologist has to depend instead on purely skeletal characters, such as the presence in the mammals of a lower jaw formed of but a single bone on each side. In the lower groups of vertebrates there are extra bones at the back of the jaw, and in the Triassic mammal-like reptiles only the last vestiges of these bones remain.

132

The Triassic is the only system in Britain above the Precambrian which is all but completely lacking in marine fossils. A few years ago a student party was visiting the Eakring oilfield in Nottinghamshire (where oil is pumped from deeply-buried Carboniferous rocks). The rocks exposed locally are Triassic, and one of the party idly hitting a road-side section with his hammer, found the first undoubted marine fossils known from the British Trias. There were several specimens of brachiopods (which are all sea animals) but of a type that is tolerant of somewhat brackish conditions. This is the only definite evidence of a sea entering the British area during Triassic times. On the continent, however, there are plenty of marine deposits of this age.

The name Trias was originally applied in Germany and referred to the three-fold division of these rocks there. The lower and upper divisions (called **Bunter** and **Keuper** respectively) are both of the continental desert type seen in Britain, and these names are used for our rocks. The middle division—the **Muschelkalk**—is marine. It yields a large but restricted fauna mostly of bivalved molluscs or 'mussels' (hence its name), though it has also some ammonites which enable it to be correlated with the Tethys. The Muschelkalk sea is known to have extended as close to Britain as Heligoland. Southwards it had only very narrow connections with the open sea of the Tethys (see Figure 37).

It is in the wonderful country of the Austrian Tyrol and the Italian Dolomites that the Trias really comes into its own. Here, thousands of feet of limestone yield corals, brachiopods, simple marine plants, ammonites and other molluscs, indicating various divisions of the Trias. Several different types of sediment can be recognized, grading from those of the northern margin of the Tethys geosyncline through those of the central deeps to those of the southern margin. All these rocks were very highly disturbed in the much later earth-movements which formed the Alps; in places the rocks of one range can be shown to have been pushed scores of kilometres over the rocks of another. It is only by sorting out the different types of Triassic (and other) deposits and where they came from, that Alpine geologists have been able to work out the very complicated structures.

The Tethyan faunas contain many new forms, and many Palaeozoic ones are significantly absent. Most of the Palaeozoic corals, brachiopods, echinoderms and arthropods had become extinct. The

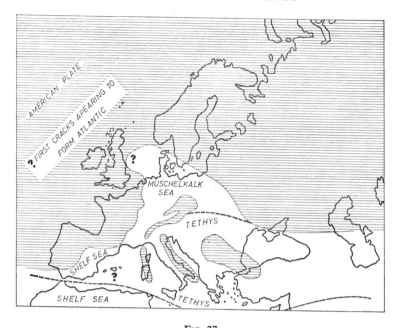

FIG. 37

Reconstruction of the supposed geography of mid-Triassic times. (Slightly simplified from *A Palaeogeographical Atlas of the British Isles* . . . by L. J. Wills, by kind permission of the author and Messrs. Blackie & Sons, Ltd.)

Palaeozoic goniatites with fairly simple partitions between the chambers were by now completely replaced by the more complicated **ammonites** (see Figure 35). Entirely new groups appeared which are still living today—for example the oysters and the modern types of corals—and this, with many other lines of evidence, justifies our regarding the Triassic as the beginning of the Mesozoic or 'Secondary' Era of life on this earth.

VII

THE SOUTHERN CONTINENT

This chapter deals with India and the continents of the southern hemisphere, which had a very different geological history from that described so far. It also discusses further the idea that the continents may in the past have drifted about on the earth's surface.

So far in this book, little has been said about the geological history of the southern hemisphere. Vast areas of dry land in this part of the world are occupied by Precambrian shields (see Figure 8). Above these ancient rocks come varied developments of Lower Palaeozoic strata, which in most cases are not very well known. Devonian and Lower Carboniferous formations have been better studied, but it is from the Upper Carboniferous onwards that this part of the world comes into its own and demands a special chapter to itself.

From Upper Carboniferous times until late in the Mesozoic, the continents of the southern hemisphere, together with India in the north, had remarkably similar geological histories. The greater part of the present land surfaces in this region was then also land—either suffering erosion or being buried under thick continental deposits. To the north they were separated from the lands of the northern hemisphere by the geosynclinal 'Tethys'. To the south there was another geosynclinal sea which impinged on the margins of the present continents. The thick continental deposits which are found throughout these regions are often called the 'Gondwana Series' after the Gond district of India where they are well developed. The remarkable similarities in the late Palaeozoic and early Mesozoic histories of India, Africa, South America, Australia and Antarctica led the great Austrian geologist Suess to coin the term **Gondwanaland** for a postulated southern continent which incorporated them all.

These similarities are the chief evidence in favour of one of the

135

most startling ideas in geology—that of **Continental Drift**—which has been discussed by geologists for many years, but has only recently led to the 'new geology' of plate tectonics. The later part of this chapter will be devoted to a discussion of the geological (as distinct from geophysical) evidence for these ideas. But first it is necessary to summarize very briefly the relevant parts of the stories of the continents that went to make up 'Gondwanaland'.

AFRICA

The keystone of the ancient continent of 'Gondwanaland' was the rigid landmass which is now Africa. Little can or need be said about the early Palaeozoic history of that continent. Apart from a few Cambrian fossils found recently in what was thought to be Pre-cambrian, there is very little record of the Lower Palaeozoic through-out the greater part of central and southern Africa. In the Union of South Africa, clearly datable sedimentation started with the **Cape Formation**—a thick succession of sandstones, quartzites and shales which rest unconformably on the older rocks. The basal beds are almost completely unfossiliferous and so cannot be dated. They in-clude beds of conglomerate in which the boulders are of all sizes and types, mixed together and sometimes scratched. Such beds are thought to be the product of ice action. Later beds of the Cape Formation contain marine fossils of Lower Devonian age, so it is thought by some that a glaciation occurred in this area very early in Devonian times or perhaps late in the Silurian. There is much better evidence of glaciations in other parts of the world and at other levels in the stratigraphical column, but these will be dealt with later.

The highest beds of the Cape Formation are of terrestrial origin and contain plant remains which date them as late Devonian to early Carboniferous in age. It will be seen that in South Africa, as elsewhere in Gondwanaland, the main divisions of stratigraphical time do not fit in very well with the standards of the northern hemis-phere. This is even more true in the strata which followed.

Above the Cape Formation in South Africa comes a very thick and important system of rocks. This is the **Karoo System** which spans the period of time represented in the north by the Upper Carboniferous, Permian, Triassic and part of the Jurassic. Four sub-

divisions are recognized within the Karoo, and the table below
shows how these are thought to correlate with the standard systems:

	South Africa	Standard Divisions
	Stormberg Series	Lower Jurassic Upper Triassic
KAROO	Beaufort Series	Lower Triassic Upper Permian
SYSTEM	Ecca Series	Lower Permian
	Dwyka Series	Upper Carboniferous

The Karoo System represents a very long period of continental
deposition. Marine sediments are only known as brief local intercalations in south-west Africa and Madagascar, and even here the
marine fauna is unusual and restricted in variety.

The most interesting deposit of the whole Karoo succession is a
thick conglomeratic division in the Dwyka Series which is nowadays
interpreted as the deposit of an ancient ice-sheet. The conglomerates
are thought to be the equivalents of the 'boulder clays' of the
Pleistocene 'Ice Age' which will be discussed in Chapter XI. They
usually consist of a great variety of unsorted, angular blocks scattered irregularly in an unbedded clayey deposit; some of the boulders
bear clear ice-scratches. In places the boulder bed rests on a **glaciated
pavement**—a smoothed and polished surface of older rock, showing
scratches produced by rock fragments being dragged along in the
ice—exactly like the rock surfaces at the sides of a modern glacier
in the Alps. The scratches are particularly useful because they indicate the direction in which the ice moved (see Figure 38).

Associated with the Dwyka 'boulder clays' are other glacial deposits, such as finely laminated shales which may have been laid
down in ice-dammed lakes. Over most of its area of deposition,

the 'boulder clay' was evidently laid down on dry land in the same way as the 'ground moraine' of a modern glacier, but in places, mainly in the south, the ice sheet apparently melted in water and the morainic deposits are fairly well bedded.

The directions of ice movements shown by the scratches indicate that it came from four main centres which succeeded each other in time from west to east. Thus there was not just one glaciation but at

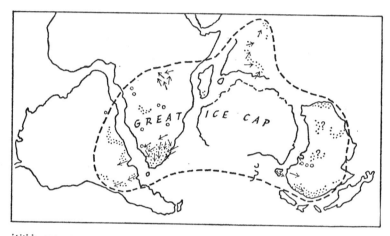

PERMO-CARBONIFEROUS GLACIAL DEPOSITS
→ SUPPOSED DIRECTION OF ICE MOVEMENT
‒ ‒ ‒ EXTREME LIMIT OF ICE
o o o PERMO-CARBONIFEROUS 'BOULDER CLAYS'

FIG. 38
The glaciation of Gondwanaland (re-drawn from *Our Wandering Continents* by A. L. du Toit, by kind permission of Messrs. Oliver & Boyd Ltd.).

least four, as in the famous 'Ice Age' of Pleistocene times in Europe and North America. Perhaps the most remarkable feature of the South African glaciation is that the main ice movement seems to have come *from the north*—that is, from the direction of the equator. There has been some dispute over the exact age of this Dwyka glaciation (and even whether it was really a glaciation at all) but the majority of workers favour the Upper Carboniferous as the most likely date.

In the strata above those of the glaciation, and in a few cases associated with it, are plant remains quite different from those of the

Upper Carboniferous and Permian in the northern hemisphere. The best known of the plants is a seed-fern called *Glossopteris* (see Figure 35) which gives its name to the flora. The 'Glossopteris Flora' is characteristic of all the lower part of the Karoo. It is very restricted in variety and this, together with the proximity of the glacial deposits, has led some palaeobotanists to suppose that it lived in arctic or sub-arctic conditions. In the Ecca Series, the plants are associated with valuable coal-seams. These are worked in the Wankie coalfield of Rhodesia, and it should be noted that these coals are later in age than the familiar Coal Measures of the British Isles. In the upper part of the Beaufort Series an important change occurs in the plants, with the incoming of typically Mesozoic forms and the disappearance of *Glossopteris* and its associates.

The most important fossils of the Karoo System are undoubtedly the vertebrates, which though usually rare, do serve to date the main divisions and to correlate them with the rest of the world. The Permian and Triassic vertebrate faunas of South Africa, together with those of Russia, also form the main source of knowledge about the critical phase in vertebrate evolution at the end of the Palaeozoic.

The only notable form in the lower part of the Karoo is the aquatic reptile *Mesosaurus* which is found in the Dwyka Series and will be mentioned again later. The Beaufort and Stormberg Series have both provided large and varied vertebrate faunas. These are mostly reptiles, of several different groups. They were nearly all large and ungainly creatures. The most important group, from the evolutionary point of view, was that which gave rise to the mammals, and which was mentioned in the last chapter.

In the top beds of the Karoo—in red deposits suggestive of arid desert conditions—the 'dinosaurs' came into their own as they did in the Jurassic of Europe and North America. Also in the topmost Stormberg Series there is evidence of widespread volcanic activity. Only tiny scattered remnants now remain of what must have been huge areas of lava and volcanic ash.

The deposition of the Karoo System in South Africa ended early in Jurassic times, and the remainder of that period over most of the continent was a time of erosion. In places, for example, near Mombasa and in Madagascar, and especially in the north, there was a marine invasion of the borders of the old landmass in Upper Jurassic times. This corresponds with what seems to have been a general expansion of the seas observed in many parts of the world at this level in the stratigraphical column.

A more extensive marine transgression, and the first to invade the southern part of Africa, occurred early in the Cretaceous. The Karoo System is overlain unconformably in many areas by marine Cretaceous strata with a thick basal conglomerate. This was one of the most important marine transgressions in the history of the earth and is discussed in Chapter VIII.

SOUTH AMERICA

Unlike South Africa, South America has a good record of geosynclinal sedimentation in the Palaeozoic. The Andes, like all mountain chains, is founded on a geosyncline, and geosynclinal sedimentation seems to have gone on in this region for a very long time. Certainly there were thick marine deposits accumulating here right till the end of the Palaeozoic. Farther east, however, away from the Andes, continental conditions set in during the Carboniferous and lasted well into the Mesozoic. The continental sediments, which were laid down in a series of large basins, are very like those of the Karoo in South Africa. Towards the end of the Carboniferous (or possibly a little earlier in parts of Argentina) there began in South America a glaciation closely comparable with that of the Dwyka Series, with advances and retreats of the ice and all the usual glacial phenomena. Higher up there are valuable coal-seams like those of the Ecca, with the same *Glossopteris* flora. Another remarkable palaeontological similarity is the occurrence in Brazil of the aquatic reptile *Mesosaurus*, which was only otherwise known in the Dwyka Series of South Africa.

After the Coal Measure episode, this region seems to have been subjected to uplift and erosion rather than subsidence and deposition, but in places further continental deposits are found with Triassic and Jurassic plants and reptiles. There was an important volcanic episode in parts of Argentina at about the same time as the Stormberg volcanic rocks were being extruded in South Africa. The period of continental sedimentation ended in South America with an unconformity under a late Mesozoic marine transgression.

INDIA

The Indian subcontinent is, in many ways, the most interesting fragment of Gondwanaland. Among other things, it is the only area where typical Gondwana rocks can be seen close up to the old dividing line of the Tethys. Continental sedimentation began here, as elsewhere, during the Upper Carboniferous. It starts with a thick boulder bed which was the subject of controversy for many years, but is now generally accepted as a 'boulder clay' laid down by ice. Above this comes the usual *Glossopteris* plants, followed, in due course, by Mesozoic types. The same floral change occurs in all parts of Gondwanaland, but it is difficult to prove that it was an evolutionary change occurring simultaneously everywhere. It might merely have been a local change in each place, brought about by the transition from damp coal swamps to more arid conditions. Higher up in the Gondwana Series of India, many Mesozoic reptiles have been found, and at the top is a thick succession of lava flows which are difficult to date accurately.

The northern edge of Gondwanaland ran roughly along the line of the Himalayas; the neighbouring continental sediments include marine intercalations. Late in Jurassic times the sea spread south over the land and laid down deposits such as the highly fossiliferous beds of Cutch in north-west India.

AUSTRALIA

The late Palaeozoic and Mesozoic rocks of Australia are as yet incompletely known, and it is difficult to summarize the evidence from widely scattered and varied basins. Geosynclinal troughs and basins bordered the West Australian shield during Devonian and Lower Carboniferous times. In those places where land deposits are known, the floras in them seem to be comparable with those in other parts of the world.

The late Palaeozoic glaciation is known in Australia in areas as far apart as Queensland and Western Australia. It seems to have been particularly effective in Victoria, where there are preserved glacial landscapes, scratched rock pavements and alternations of

141

glacial and melt-water deposits. There is evidence to suggest that the glaciation did not reach Australia until Permian times, and may in fact have moved progressively eastwards with time around the southern hemisphere, starting in South America and finishing in Australia.

At about the same level as the glaciation, the *Glossopteris* flora appeared, and the glacial deposits are followed by valuable Permian Coal Measures. In certain areas, either alternating with the Coal Measures, or completely replacing them, there are marine Permian deposits. Some of these have a restricted fauna like one found in Madagascar, but elsewhere there is a large and varied marine fauna which can be correlated with the northern hemisphere.

The subsidence which accommodated the Permian sediments was followed by uplift and erosion. The Triassic and Jurassic Systems are often completely absent, and marine Cretaceous sediments in some places rest directly on the Permian. Where the Trias is present, it is a land deposit and thick sandstones and conglomerates are the rule. These may have been laid down under arid conditions, but there are plant remains in places which are like those of the upper part of the Karoo System in Africa.

Continental deposition probably continued into the Jurassic and at about the junction of the two systems there was some volcanic activity, though not great outpourings of lava.

NEW ZEALAND

New Zealand hardly enters into the Gondwanaland picture at all, and its place in the plate tectonic story is not yet entirely clear. It may be that it came within the southern geosyncline which bordered the other regions so far discussed. Many thousands of feet of geosynclinal deposits are known, with associated pillow lavas and ashes, which are referred to the Carboniferous and Permian on the basis of occasional fossils. The late Palaeozoic glaciation of Gondwanaland has not been recognized. The Triassic rocks which follow seem to be wholly marine with normal faunas such as ammonites and brachiopods, and the Jurassic includes records of both marine and non-marine conditions. But all the earlier rocks of New Zealand are considerably disturbed, altered and confused by an orogeny which occurred at the end of the Jurassic. There are thousands of feet of sediment forming much of the mountain backbone of the country which have not yet been sorted out. It is probable that we have here a story as complicated and as difficult to unravel as that of the Alps, which will be discussed in Chapter X.

Above this confusion come highly fossiliferous Cretaceous sediments which have not suffered in the same way.

ANTARCTICA

Inaccessibility and a thick ice cover limit our geological knowledge of the Antarctic continent. Land conditions with thick sedimentation seem to have prevailed at least from Upper Devonian times, possibly as late as the Cretaceous. Volcanic activity occurred at times. Thick coal seams, including anthracite, are known but are never likely to be exploited. Before this is technically possible, such crude forms of fuel will long be outdated. Plant remains of various ages are abundant. Dr. Edmund Wilson, the naturalist of Scott's tragic last expedition, collected the first examples of *Glossopteris* known from this continent and several fine examples were brought back by the recent British Trans-Antarctic expedition. The conditions which permitted such luxuriant plant growth must clearly have differed very much from the conditions prevailing today.

South African geologists have seen many resemblances between Antarctic geology and their own, and this was encouraged by the recent discovery of *Mesosaurus* in Antarctica. The Falkland Islands, off South America resemble South Africa in remarkable detail. Charles Darwin made the first large collection of fossils here during his famous *Beagle* voyage. The Permo-Carboniferous glaciation is well seen in the Falkland Islands, and has now been recognized in Antarctica, where it is, however, largely concealed by a glaciation of much more recent date.

CONTINENTAL DRIFT

From the above accounts it is obvious that there are remarkably close resemblances in the geology of the different parts of 'Gondwanaland'. This is particularly true for that part of the record between the Upper Carboniferous and the end of the Jurassic. These resemblances are not only in the sediments, but also in the volcanic rocks, the fossils and certain of the structures. Such resemblances and other lines of evidence which will be considered later, are the chief points that led to the theory of 'continental drift'.

This idea originated in the early years of the present century, though its basic inspiration may be much older. In 1910, F. B. Taylor suggested a movement of land away from the poles as a result of the tidal effects produced when the moon was first captured by the earth. Others have put forward the idea that the moon was torn from the earth quite late in geological history, leaving a great 'lunar scar' which is now the Pacific Ocean. The tendency of the continents to drift into the gap left by the moon then produced the phenomena of continental drift. Such sensational mechanisms are not nowadays considered at all seriously. Apart from anything else, it is most unlikely that the life on earth would have survived such a catastrophe.

The man chiefly responsible for the fully elaborated theory of continental drift was Alfred Wegener, professor of meteorology and geophysics at Graz, and it is often known as **'Wegener's Hypothesis'**. Shortly before the First World War he published his ideas of a single great landmass which began to break up during Mesozoic times. One of the points which had struck him most forcibly was the remarkable 'mirror-images' provided by the opposing coastlines of South America and Africa. In his enthusiasm he certainly fitted them together far too closely, without making any allowance for continental shelves and ridges which are only covered with comparatively shallow water. Once fitted together, however, they solve a number of problems. The resemblances in the Upper Palaeozoic and Mesozoic successions become a lot easier to understand if the two land masses were at the time much closer together and continuous. It also explains the presence on both sides of the South Atlantic of the *Glossopteris* flora and the reptile *Mesosaurus*, neither of which could have migrated across deep water. Similarly, fold structures in Africa can be demonstrated as continuing along the same lines in South America.

Wegener was a meteorologist, and it was in the explanation of past climates that he found his drifting theories most useful. The outstanding climatic event in the southern hemisphere during the period under consideration was the glaciation, and this only really becomes understandable if we allow not only 'drifting' but also the associated hypothesis of 'polar wandering'. This postulates that the poles have moved considerable distances relative to the continental masses during geological time. The most outstanding protagonist of Wegener's hypothesis was the South African geologist A. L. du Toit, and the validity of the theories has become almost a matter of faith among geologists from that part of the world.

Du Toit developed and expounded the theory of continental drifting in a masterly way in his book *Our Wandering Continents*, supported with overwhelming masses of evidence from every aspect of geology. He postulated not one 'ancestral' continent as did Wegener, but two. These were 'Laurasia' in the north, consisting of

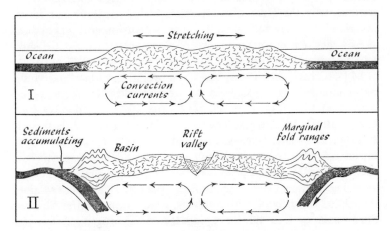

FIG. 39
First stages in the break-up of a continent by convection currents.

what is now North America, Greenland, Europe and the greater part of Asia, and 'Gondwanaland' in the south as already defined. These two were separated by the geosynclinal 'Tethys' and began to break up, through an internal mechanism, in late Mesozoic times. The running together of parts of the two continents then produced the Alpine and Himalayan mountain ranges, whilst the other fragments moved into the Pacific and developed buckles along their leading edges which are now ranges such as the Andes of South America. It should be noted that the main structures around the Pacific are all parallel to the coasts, whilst those around the Atlantic strike the coast at an angle.

In his book, du Toit reconstructed 'Gondwanaland' as shown in

Figure 38. On this figure there is also shown the maximum extent of the Permo-Carboniferous ice sheets and the supposed directions of ice movement. It will be seen that these distributions and directions become much more logical if the fragments of Gondwanaland are reassembled in this way and the South Pole is placed east of South Africa.

In the same way, a closing up of the North Atlantic, bringing North America, Greenland and western Europe close together, solves many stratigraphical problems. It explains anomalies in trilobite distribution in the Cambrian, it explains the remarkable resemblances in the Coal Measures of North America and Europe, it explains why the Hercynian structures drive headlong out to sea in Ireland, Brittany and Spain. It explains many other things besides.

As a mechanism for the break-up of contents, du Toit cited slow-moving convection currents in the semi-liquid internal layers of the crust (see Figure 39). These, operating over a very long time would, he maintained, have produced the desired effect. It should be explained at this point that the continents are essentially layers of comparatively light, acidic material resting on a denser, basic substratum. The acidic layer is, to all intents and purposes, missing in the oceanic regions. It was difficult to understand, however, why convection currents did not produce a break-up of the continents earlier in geological time. Conversely, though these ideas may have explained the Alpine Orogeny of Tertiary times, which produced the Alps and the Himalayas, they did not explain the earlier orogenies. Now that we have the general theory of plate tectonics, the picture has become much clearer. We can now see that the continents probably split up and came together in different ways several times during the history of the earth. We still depend on an internal mechanism of convection currents, which is not altogether satisfactory, but the fact that we cannot produce a satisfactory explanation of the mechanism is not in itself a good argument, for there are many major geological events such as glaciations, which certainly happened but which we cannot as yet completely explain in mechanical terms.

If the continents and oceans did drift apart on their plates, like solid ice moving in a glacier, then this must have involved considerable tension, especially in those regions where the actual break-up occurred. In this connection we can point to the great system of **rift valleys** which extend north to south from the Middle East far down into Africa and down the Red Sea. These rift valleys consist of very long but comparatively narrow tracts of country which have dropped down between parallel faults, and would seem to be good evidence for tension. Many other rift systems exist around the world—for example along the Rhine and the Rhône and along our own Vale of Severn. We can now see quite clearly that the Red Sea was torn open in this way. In the same way, the tearing open of the North Atlantic is thought to have produced the great outpourings of lava which occurred in that region in early Tertiary times.

As has already been mentioned, du Toit produced many different arguments to support the drifting hypothesis. Apart from purely geological evidence, he quoted several examples of modern animal and plant distributions which seem to require former connections between the continents. One of these interesting modern distributions is shown in Figure 40. It shows the areas of occurrence of different

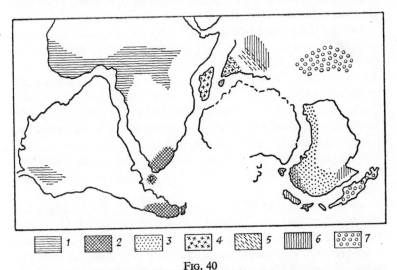

FIG. 40

Map showing the distribution of certain genera of living 'rain-worms' (family MEGASCOLECIDAE). 1: *Dichogaster*; 2: *Chilota*; 3: *Megascolex*; 4: *Howascolex*; 5: *Octochaetus*; 6: *Perionyx*; 7: *Pheretima* (re-drawn from *Our Wandering Continents* by A. L. du Toit, by kind permission of Messrs. Oliver & Boyd Ltd.).

genera of rain-worms, which are terrestrial forms sometimes of very large size. When a genus or a species is found in two widely separated places, it must presumably have migrated from one to the other (or to both from somewhere else) at some time in the geological past.

Some biologists have postulated long land bridges between the continents along which land animals and plants migrated. But so many land bridges are required at so many different periods and in so many different directions that they become even more difficult to accept than 'drifting'. Most workers now postulate not just a single drifting apart, but a movement of the continents to and fro.

Another way of explaining anomalous distributions was the process of 'rafting'. The occurrence of certain small rodents on both sides of the South Atlantic might be explained by a single female being accidentally carried across on a floating log by ocean currents in a 'Kon-Tiki' manner. On arrival at her destination she might then have produced a litter and thence a new race. So many coincidences are required for such an accident to be successful that it seems hardly worth considering. But with all the vast resources of geological time to call upon it is not so impossible. Given enough time a possibility becomes a probability and eventually even a certainty.

Perhaps the most far-fetched and amusing argument of all was provided by the remarkable habits of the eel. Every autumn adult eels come down the rivers of Europe and swim right across the Atlantic to spawn near Bermuda.[1] Each spring the newly-born elvers return to the same rivers. It was once suggested, as an explanation of this remarkable migration, that it started when the Atlantic was little more than a narrow crack between the continents. Then, as the two sides drifted slowly apart, the eels had to swim a few extra inches every year, but their racial memory never noticed the difference, and so the migration eventually reached its present marathon proportions!

An interesting line of evidence which has appeared some years ago was the discovery by geophysicists of the original magnetic properties of rocks. For reasons that are not yet fully understood, there are certain sedimentary and volcanic rocks, usually rich in iron, which still retain the magnetic properties which they had when they were first laid down or extruded. These properties give a clear indication of the magnetic field of the earth at the same past moment in time. The results so obtained are often very surprising. Thus, studies of the Triassic rocks of Britain indicate a magnetic field rotated about 40° to the west of where it is now. This shows that the

---

[1] Since the above was written, part of this theory has been challenged by Dr. D. W. Tucker, but it is such a nice story it is a pity to leave it out.

magnetic poles of the earth were then in a different position relative to the part of the surface which we now call Britain.

Similarly, studies of basalt lava flows in India (i.e. north of the equator) show a general magnetic field dipping southwards towards a pole in that direction. This work is quite separate from that referred to in Chapter I in which the periodic reversal of the earth's magnetic field was considered. But the evidence provided of the direction and of the poles for any given land area provides us with a wonderful means of determining the latitude and orientation of our land masses at given points in past time, though not, unfortunately, of their longitude. There have been a great number of contradictory conclusions reached on the basis of this work, though in general and certainly in the later part of the earth's history they seem to fit in reasonably well with the ideas of plate tectonics. Farther back in time (i.e. before the Mesozoic) the results seem to be more doubtful.

The author's own private prejudices are towards the evidence provided by palaeontology. Much fossil evidence (such as *Glossopteris* and *Mesosaurus*) can be cited to support the conclusions of the geophysicists, but there is a fair amount of fossil data which seems to argue against polar wandering if not against drifting. This applies particularly to the elucidation of past climatic belts by the study of fossil distributions. Various groups of plants, molluscs, brachiopods and foraminifera have been shown to be distributed in zones parallel to the present equator. This does not eliminate drifting, but it does imply a fair constancy in the position of the poles and not much rotation of the continents. The general picture seems to be that in the geological past, the tropics have usually been much wider than they are at present, but they have been more or less parallel to the present lines of latitude, at least in the later part of the story.

The Gondwanaland picture is therefore a very clear one, but more recent evidence has shown an equally clear picture of a northern continent, or 'Laurasia'. Wegener originally postulated an original single continent of 'Pangaea' which broke up to form his other two daughter continents. The picture that seems to be emerging now seems, to the author at least, to take us back more and more to this original mother continent, which has repeatedly broken up and come together again in the course of geological history.

VIII

THE LONG QUIET EPISODE

This chapter deals with the long quiet episode, without volcanic activity or major earth movements, which followed the Hercynian orogeny and the New Red Sandstone deserts. This lasted through the Rhaetian Age and the Jurassic and Cretaceous Periods to the end of the Mesozoic.

THE RHAETIAN AGE

About seven miles west-south-west of Gloucester is the small village of Westbury-on-Severn. Here the Severn sweeps round a left-hand bend and the river cuts into its northern bank revealing a fine geological section. Downstream, towards Newnham, are the monotonous red dust deposits of the Keuper Marl, dipping gently to the south-east. Towards the top of this, green bands appear alternating with the red, and the uppermost beds are wholly green. This transition has been interpreted as representing the change-over from the oxidizing conditions of the Keuper desert to the reducing conditions of stagnant water deposition. Above the green marls at Garden Cliff, Westbury, come a remarkable series of beds known as the Rhaetian. A similar section is seen downstream at Aust Cliff—the site for the Severn bridge—and is shown in Plate 14.

The Rhaetian Series is named after the Rhaetian Alps on the borders of Austria and Switzerland, where strata of this age are thickly developed. On the continent, the Rhaetian is properly regarded as the top division of the Trias. In Britain it is linked with the Jurassic because it marks the start of the long cycle of marine sedimentation which only ended at the close of the Jurassic Period.

The Rhaetian sea spread up over northern Europe from the Alps, quietly flooding the low-lying desert plain of the Keuper Marl. So gentle was this marine transgression that no sharp break can be seen between the continental and the marine deposits. In places, where upland areas of older rocks stood up from the desert, the Rhaetian sea flowed round the sides and left them as islands. There was a whole archipelago of islands around what is now the Severn estuary. The largest of these was the Mendip ridge, which was not finally flooded until Middle Jurassic times (see Figure 44). The greater part of Wales is thought by some geologists to have been a land mass at this time, though others (including the author) now think that it was open sea. One of the most remarkable pieces of evidence in favour of this contention was the borehole on Mochras 'shell island' south of Harlech in North Wales where—within a stone's throw of the oldest Cambrian—the borehole revealed the thickest Lower Jurassic known anywhere in the British Isles. There certainly was a large island, however, extending from the London area eastwards into northern France.

The Rhaetian sea of northern Europe was unusual and probably rather unhealthy, in some ways rather like the Muschelkalk sea of the Middle Trias. It connected with the 'Tethys' through narrow straits and only a limited number of species populated its restricted waters.

The ammonites, which had evolved to an amazing degree of variety and abundance earlier in the Trias, reached a climax in their history during the Rhaetian. They very nearly became extinct; only a single small family lived on in the 'Tethys'. None came into the shallow Rhaetian sea of the north. Later this small family was to give rise to another great evolutionary 'explosion', producing all the many and varied ammonites of the Jurassic and Cretaceous. It is, perhaps, the only example of an evolutionary lineage which, through some combination of inheritance and opportunity, had a second chance in evolution. This was fortunate for Mesozoic stratigraphers, as the ammonites are their chief arbiters of correlation.

The fauna of the Rhaetian sea which spread over desert Britain was restricted and poorly developed. It consisted mainly of small lamellibranchs. This may have been because the sea-floor was too unhealthy for other forms of life. The first sediments to accumulate in this sea were black sulphurous muds, which are seen today as black shales full of crystals of the iron sulphide, pyrites. At or near the bottom of the shales is the most remarkable of all the Rhaetian deposits—the 'Rhaetic Bone' Bed—a thin layer packed with bones, teeth and excreta of reptiles and fish. This is probably a 'winnowed' deposit

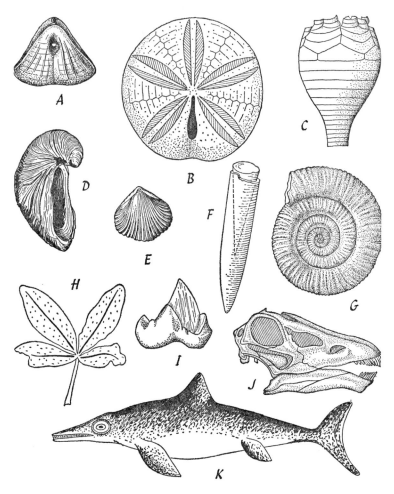

FIG. 41

Jurassic fossils. *A*: aberrant brachiopod from the Alps, $\times \frac{2}{3}$; *B*: large sea-urchin or echinoid, $\times \frac{1}{3}$; *C*: part of stem and cup of an unusual crinoid (after Institute of Geological Sciences), $\times \frac{1}{2}$; *D*: oyster with strongly incurved apex, $\times \frac{1}{2}$; *E*: brachiopod, $\times \frac{1}{2}$; *F*: belemnite counterweight or guard, with part of conical shell in which the animal lived (after Davies), $\times \frac{1}{2}$; *G*: ammonite (after Institute of Geological Sciences), $\times \frac{7}{8}$; *H*: leaf of seed-bearing plant (after Seward), $\times \frac{1}{2}$; *I*: early mammalian tooth from Swanage (after Simpson), $\times 14$; *J*: skull of a large herbivorous dinosaur $\times \frac{1}{14}$; *K*: reconstruction of a marine reptile, $\times \frac{1}{24}$.

like the Ludlow Bone Bed discussed in Chapter III, and it is perhaps significant that both occur at the point of transition from
continental to marine conditions. Besides fish there are representatives of the various groups of reptiles which reverted to an aquatic
mode of life at this time; there are also examples of the mammal-like
reptiles. The teeth of one of these have been found with other fossils
in fissures in the top of the Carboniferous Limestone near Frome in
Somerset. They are thought to have accumulated from a land surface
on one of the islands of the neighbourhood.

The black shales of the Rhaetian extend as far as the Hebrides and
north-east Ireland. Later beds in the west of England include thin
sandstones and unusual chemically-deposited limestones. One of
these—known as the 'Landscape Marble'—has the appearance, when
polished, of a landscape with buildings and trees. It has been suggested that this effect was due to bubbles of gas passing up through
fetid calcareous muds on the sea-floor, but it now seems more likely
that it was formed by algae.

THE JURASSIC PERIOD

At Pinhay Bay, west of Lyme Regis on the Dorset coast, one can
see the white limestone at the top of the Rhaetian dipping gently eastwards under the lowest division of the Jurassic. The latter continues,
in a wonderful series of cliff sections, all along the Dorset coast
nearly to the Hampshire border. The Jurassic sea followed on the
Rhaetian without a break (see Figure 43).

The Jurassic rocks of Britain may be said to be the cradle of
geology, for William Smith—the 'father of English geology'—was
born on the Jurassic and first worked and studied amongst the rocks
of the period. He was the son of an Oxfordshire blacksmith, and
become a surveyor and civil engineer during the great days of canal
building at the end of the eighteenth century. He later travelled all
over England and produced the first real geological map, but it was
in the Jurassic that he first recognized the fundamental principles of
the succession of bedded rocks. He was fortunate in that the rocks on
which he found himself were varied, highly fossiliferous (with the
most useful kinds of fossils, see Figure 41) and little disturbed by
earth movements. Equivalent strata, in a similar state, occur in
south-west Germany, and there saw the start of modern strati-

graphical palaeontology with the classic German workers of the nineteenth century, Quenstedt and his pupil Oppel.

About sixty faunal zones are now recognized in the Jurassic and very many more sub-zones; most of these are recognized by and named after, species of ammonites. The Jurassic rocks of north-west Europe fall naturally into three main divisions, Lower (or **Lias**), Middle and Upper. In Germany, geologists still use Quenstedt's terms Black Jura, Brown Jura and White Jura respectively, which are surprisingly apt for the British succession as well.

The Lower Jurassic or Lias extends in a broad sweep from the Dorset coast to the Yorkshire coast, with separate patches or outliers in Glamorgan, Shropshire, Cumberland, the Inner Hebrides, north-east Scotland and north-east Ireland (see Figure 42). The Lias is largely dark clays and so forms broad clay vales such as those of Gloucester and Evesham. On the Dorset coast it follows on the Rhaetian west of Lyme Regis and its various divisions are seen along the coast as far as Burton Bradstock. The fossils of the Lias have long been famous, and until quite recently local collectors at Lyme Regis in Dorset and Whitby in Yorkshire have earned a living selling fossils to holiday visitors. At Whitby, the abundant Liassic ammonites are known as 'snake-stones' and were thought to have been snakes petrified by the local saint—St. Hilda. One of the dealers at Lyme Regis was Mary Anning who, as a young girl at the beginning of the nineteenth century, found the first British examples of the large marine reptiles for which the Lias is famous. She supplied specimens to many of the leading geologists of her day.

With the ammonites and reptiles are many other types of fossils, but mainly those kinds which thrived in muddy seas. Thus there are great numbers of oysters, which provided one of the author's predecessors at Swansea, Sir Arthur Trueman, with a classic example of an evolutionary lineage, from flat to tightly coiled forms. Unfortunately it now seems that Sir Arthur got it wrong, but it was a very good idea!

The only major interuption in the muddy sedimentation of the Lias was in the middle of the epoch, when the sea became shallower and the sea-water became rich in iron. A series of iron-rich limestones were deposited, which around Banbury in Oxfordshire, in Leicestershire, Lincolnshire and east Yorkshire are sufficiently rich to be worked as an iron-ore. The Cleveland Ironstone of Yorkshire was until recently the most important source of iron-ore in Britain.

Being a harder bed than the clays above and below, the Middle Lias iron-stone often forms a strong scenic feature. A good example of this is the steep escarpment at Edge Hill in Warwickshire, down which Prince Rupert charged with his cavalry in the famous Civil War battle. Between here and Banbury the iron-stone was still worked on a huge scale until recently.

Similar bedded iron-stones are known in various parts of the Jurassic in other areas. There are the extensive workings of the Lower Liassic Frodingham Iron-stone around Scunthorpe in Lincolnshire. In the island of Raasay in the Inner Hebrides, there is a thin Upper Liassic iron-stone which was discovered by a geologist and worked during the First World War. In the Middle Jurassic there are the biggest workings of all—huge trenches miles in length around Kettering, Wellingborough and Corby—for the extraction of the Northampton Sand Iron-stone. In the Upper Jurassic there are only minor iron-stones—for example at Abbotsbury in Dorset and at Westbury in Wiltshire.

The origin of these iron-stones remains a considerable mystery. They are called 'bedded' iron-stones since they occur in regular beds, like other sedimentary deposits, and are quite unlike other bodies of iron-ore which cut across and replace the 'country' rocks. Nothing like the deposition of iron-stone is known to be going on in modern seas, so the bedded iron-stones are one geological phenomenon which we cannot interpret by reference to the present.

Apart from the iron-stones in the Lias, the clays also are of considerable value and are dug for brick-making in innumerable pits from Dorset to Yorkshire. At Battledown, on the outskirts of Cheltenham, for example, the Lias clay is said to have been worked for bricks and tiles since the days of King Alfred. There are some very thin limestones in the Lias as well, especially near the base and near the old shorelines, and these were worked in the past for lime.

Just below the top of the Lias in Gloucestershire is a bright yellow sandstone known as the 'Cotteswold Sands'. Farther south, in Somerset, are the very similar 'Midford Sands' and, beyond the Mendips, the 'Yeovil Sands'. Finally, on the Dorset coast at Burton Bradstock are the steep impressive cliffs of yellow 'Bridport Sands' with harder bands weathered out as ledges. All these sands are very much alike in appearance, and past geologists might logically be blamed for their parochialism in giving them all different names. However, detailed studies of the fossils in these sands have shown

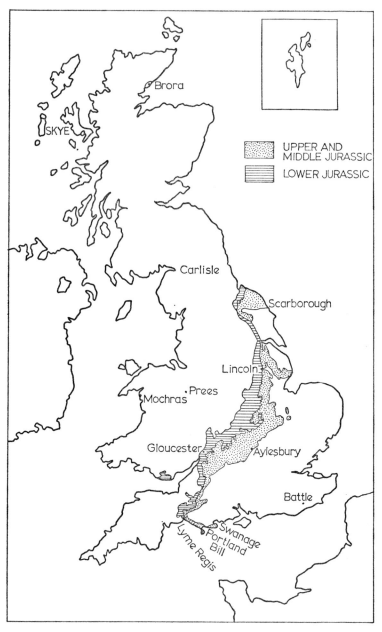

Fig. 42
Map showing the main outcrops of Jurassic rocks in Britain.

that they are not everywhere of the same age. We have here, in fact, a classic example of the phenomenon of **diachronism** which bedevils stratigraphical geology (see Figure 45). Whereas the Cotteswold Sands in Gloucestershire are wholly within the Upper Lias, the seemingly equivalent Bridport Sands in Dorset largely belong to the Middle Jurassic. The sands in Somerset come between the others both in space and time. In other words there was a moving belt of sand deposition which started in Gloucestershire during Upper Lias times and moved slowly southwards through thousands and thousands of years.

The Bridport Sands brings us to the Middle Jurassic, which, in Britain consists mainly of shallow-water limestones. William Smith in his early excursions around Bath, recognized subdivisions in what is known as the **'Great'** or **'Bath Oolite'**. This is a thick series of limestones which forms the local hills. Below this series he found a minor limestone development which is therefore called the **'Inferior Oolite'**. Higher up he found a thin but persistent limestone which he

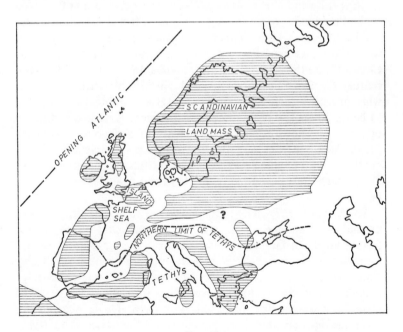

FIG. 43
Reconstruction of the supposed geography of Europe during early Jurassic times

called the 'Cornbrash' because it gives rise to a rubbly or 'brashy' soil which is very suitable for growing corn. These three divisions, the Inferior Oolite Series, the Great Oolite Series and the Cornbrash form the Middle Jurassic as it is developed in England. The term **oolite** refers to a particular type of limestone made up of tiny spheres which closely resemble the roe or eggs (Greek: *oon*) of fish. These little spheres consist of concentric layers of calcium carbonate around a nucleus. They appear to have formed in shallow, agitated water.

The Middle Jurassic oolitic limestones sometimes split easily into blocks and so have been quarried in many places for building stone. The 'Bath Stone' is particularly famous in this connection and besides its local use has been quarried for such buildings as the Duke of Wellington's house at Hyde Park Corner. The same strata in Oxfordshire have been used for centuries to provide the warm stone of the Oxford colleges, though certain of the beds do not react very well to a modern car factory atmosphere. One of the beds of the Great Oolite in Oxfordshire is the so-called 'Stonesfield Slate' and the tips from the mines where it was once worked can still be seen around the village of that name. It is not in fact a slate, but a well-bedded limestone which splits to give thin slabs that have been used for centuries as a roofing material. They can be seen all over the Cotswolds on the roofs of the older houses. In the middle of the last century the Stonesfield Slate aroused considerable scientific interest when among its large and varied fauna there was found what proved to be the lower jaw of an undoubted mammal. This was and is one of the oldest known mammals in the world. The deposit cannot have accumulated far from the shore, for mixed up with its marine fossils are the remains of plants and dinosaurs (besides the famous mammal) all of which must have lived on dry land and floated out to sea after death.

The Inferior Oolite limestones, which are sufficiently insignificant around Bath to deserve their name, become very much thicker farther north. In Gloucestershire they form the magnificent escarpment of the Cotswolds above the Lias lowlands of the Severn valley These reach their finest development between Wootton-under-Edge and Broadway, and include the only hills in south-east England to exceed a thousand feet in height. The Great Oolite quarries are down the dip slope, but along the escarpment centuries of quarrying have produced high faces of Inferior Oolite such as those of Cleeve Hill and Leckhampton Hill above Cheltenham, which have the appearance of natural cliffs. Certain of the limestone beds are grey or white in colour, but the majority have a wonderful golden-brown appearance which has been called 'the sunlight in the stone'. As a building stone, the Inferior Oolite is seen at its best, combined with Stonesfield Slate roofs, in villages such as Painswick, Chipping Camden and much-visited Broadway and Bourton-on-the-Water.

Most of the oolites are richly fossiliferous, with brachiopods, sea-urchins and certain molluscs, though there are very few ammonites, which apparently did not like this type of environment. Corals are found in some beds in the Cotswolds and these are thought to be debris from a line of reefs that grew somewhere in the vicinity. The Mendip archipelago was probably drowned for the first time in Middle Jurassic times. South of here the limestones become very thin and are insignificant on the Dorset coast. Clays are developed in the Great Oolite Series and one of the most fossiliferous localities in England is at this level at Langton Herring near Weymouth, where brachiopods could easily be collected by the hundred.

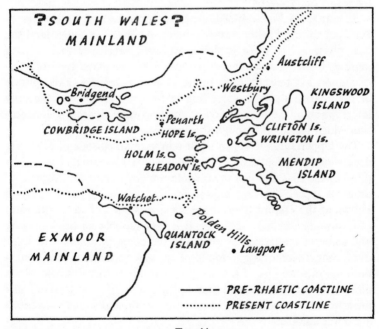

FIG. 44

Sketch-map of St. David's archipelago at the beginning of Rhaetian times (after Arkell—from earlier authors).

THE LONG QUIET EPISODE

North of the Cotswolds, the Middle Jurassic rocks change markedly in character. The Northampton Sand Iron-stone has already been mentioned. This is, at present, Britain's most valuable source of iron-ore and comes within the Inferior Oolite Series. One thick limestone formation continues somewhat as before to form the remarkably straight feature of the 'Lincolnshire Wall' which passes through the city of Lincoln and raises to heaven its dominating cathedral. Besides the usual and unusual marine sediments, there came into the Middle Jurassic hereabouts, a series of deltaic deposits laid down by a river system flowing from the north. These are better seen in Yorkshire where both the Inferior Oolite and the Great Oolite take on a form rather like the Carboniferous Coal Measures. Cycles of sedimentation are seen on the Yorkshire coast, ranging from normal marine limestones to seat-earths and poor coals. A particularly fine 'washout' can be seen in the 'Spion Cop' car park at Whitby. The coastal strata between here and Runswick Bay are famous for their plant remains and much of our knowledge of the Jurassic flora comes from this one locality.

Similar Coal Measure sedimentation prevailed on both sides of Scotland during Middle Jurassic times. At Brora in Sutherland the coal phase of the cycle is sufficiently well developed to be worth mining. Even farther afield—in Denmark—the same type of sedimentation has recently been found in deep boreholes. This episode of non-marine deposits spreading in from the north was the only interruption of the long period of shallow-water, marine sedimentation which lasted all through the Jurassic.

The Upper Jurassic began in Britain with an episode of relatively deep-water sedimentation. The muds which accumulated were like those of the Lias and like them contain many ammonites, though of different types. The first muddy division of the Upper Jurassic is known as the **Oxford Clay**, though it is difficult to find an exposure of it nowadays around that city. It is very suitable for brick-making and some of the largest brick-pits in the country, notably the huge ones near Peterborough, are dug in this formation. The Peterborough pits are one of the few places in the British Jurassic where dinosaur remains have been found. The bodies of these land animals must have been washed out to sea after death, and so their skeletons are never complete.

In other parts of the world, for example in west-central U.S.A., there are thick continental deposits of Jurassic age and in these the remains of giant reptiles are often remarkably well preserved. The dinosaurs are probably the best-known fossils so far as the general public is concerned, but the average geologist is never likely to find one outside a museum. Their evolutionary story is fascinating, but founded as it is on a limited number of isolated specimens, it cannot be examined in anything like the scientific detail of those of humbler creatures.

The **dinosaurs** arose from Triassic ancestors and must have dominated the land surface during Jurassic and Cretaceous times. The popular name 'dinosaur' covers several different families which evolved independently. Their ancestors walked on their hind legs only, and many of the later forms retained this habit. Some were plant feeders, browsing on the ferns, conifers and other primitive seed-bearing plants which covered the land. Others were flesh-eaters and lived on their vegetarian contemporaries. Some groups of dinosaurs reverted to walking on four legs, though the front pair are

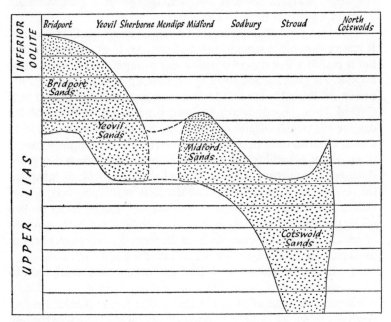

FIG. 45

Diachronism in Lower and Middle Jurassic sands. The diagram shows how the deposition of the sands cuts across the time planes (shown horizontally) from place to place, and becomes progressively later towards the south (after Arkell).

often much reduced. These included the giant forms, up to 100 feet in length, which probably lived partly submerged in rivers or lakes. These were plant feeders as were the quadrupedal armoured dinosaurs, which developed a fantastic variety of plates, frills, spikes and bosses to protect themselves from their carnivorous cousins. Though we usually think of the dinosaurs as giant forms, there were also quite small lightly-built types. In other words, the land reptiles were adapted for many different modes of life and so dominated the world not simply by their bulk, but also by their variety.

At the same time as the dinosaurs held the land, so other reptiles dominated the sea and sky. The marine reptiles of Lyme Regis have already been mentioned; these persisted in varying form right through to the end of the Mesozoic. Other reptiles (such as *Mesosaurus*) had returned to the swimming habits of their ancestors in freshwater. The **pterodactyls** were reptiles which went to the opposite extreme and were able to fly in an unwieldly manner. Their forelimbs are highly adapted for this purpose, though in a fundamentally different way from those of the birds.

Above the Oxford Clay comes a formation known as the **Corallian**. It is very variable in composition and includes both sands and clays, but is mainly a limestone formation abounding in corals, as its name implies. At places along the English outcrop actual coral-reefs are preserved. Probably the best-known example is at Cumnor Hill near Oxford. Here, shortly before the war, suburban growth caused the digging of extensive drains and foundations. These revealed the local Corallian coral-reefs in considerable detail. The reefs proper consist of masses of branching corals still in the position in which they grew. Associated with them are molluscs and sea-urchins which inhabited the interstices of the reefs. Around the reefs are sloping beds of reef debris. These consist of fragments of the reef and its fauna broken off by the waves; the corals in them are worn and not in the position of growth. The farther one goes from the reefs, the more finely broken up is the reef material, until it passes into the finest coral dust.

Also in the Corallian near Oxford are clay deposits which seem to be the sediment of river distributaries cutting through the reef belt. A comparable state of affairs is seen in the Great Barrier Reef of Australia. Here there are gaps opposite the mouths of rivers because the presence of mud in the water inhibits coral growth. Modern corals are very fussy animals and only tolerate a limited range of environment. Reef corals are unable to live below certain water temperatures. This has led to much speculation about fossil forms and the conditions implied by their presence in a rock; such theorizing is interesting but may be unscientific. It was suggested in the last chapter that the present tropical belt may be much narrower than it has been through much of geological time. Coral reefs have commonly flourished as far north as Britain, thick forests have grown in Greenland and Antarctica and hot deserts have extended across northern Europe.

The return of muddy conditions after the Corallian produced the deposit known as the **Kimmeridge Clay** named after a village on the Dorset coast. It is like the Oxford Clay but much thicker, and in places passes into shallow-water sands. In the narrow strip of Jurassic rocks along the coast of Sutherland (see Figure 42), the Kimmeridge Clay is remarkable in that it contains many great boulders of Old Red Sandstone mixed up with the bedded clays. Huge boulders like this are quite out of place in such a setting, and the section has caused much discussion. The latest interpretation is that the blocks fell from a submarine escarpment during earthquakes. The coast was not far away to the west, for the greater part of Scotland is thought to have stood up as an island during Jurassic times (see Figure 43) and the sea only impinged on its outermost edges.

At about this time in what is now Bavaria, there was being laid down a remarkable deposit known as the **Solenhofen Stone**. This is an extremely fine-grained limestone which was apparently deposited in a peaceful lagoon; it was much quarried in the past for use as a lithographic stone in old-fashioned printing. In this limestone the most fragile organisms such as jelly-fish are preserved as impressions, and conditions were so quiet that the skeletons of fish and other organisms did not fall apart. The most remarkable series of discoveries made in the Solenhofen Stone started with that of a single feather in 1860. This was exciting because a feather could only mean a bird, and the history of the birds was then virtually unknown. Their mode of life and the fragility of their skeletons make birds almost the rarest of fossils. The discovery of the feather was followed by the finding of two almost complete bird skeletons in 1861 and 1877. One of these specimens is in Berlin, the other in the Natural History Museum in London. They are certainly birds with feathered wings formed in the normal way, but they also have many reptilian features such as teeth and long tails. In fact they are the answer to the Victorian's dream of a 'Missing Link', in this case a link between the reptiles and the birds. 'Missing Links' are in practice only 'Will o' the Wisps', for every one that is found requires more links to join it into the chain; in a sense every single fossil is a 'Missing Link' somewhere in the complex pattern of evolution.

Returning once more to Britain; shallow-water conditions began to appear in several places in Kimmeridge times and from then on the Jurassic story is one of a sea retreating southwards from the British Isles. The **Portland Beds** which followed are only found in southern England. Their type locality is Portland Bill—the massive promontory just west of Weymouth which is all but an island. The characteristic outline of Portland Bill can be recognized for miles along the coast. At the base of the promontory the Kimmeridge Clay forms a gentle slope, then comes the Portland Sand forming a steeper slope and at the top are sheer cliffs of Portland Stone.

The Portland Stone consists of a varied series of pale limestones, all of which yield rich marine faunas, including giant ammonites which are often seen decorating local gardens. The more massive beds provide the famous Portland building stone, which is used in most of the larger public buildings in London and has been exported to many parts of the world. It was the stone Sir Christopher Wren chose for St. Paul's and since then it has been used for scores of pseudo-classical façades. The Portland Beds provide much of the glorious coastal (and military) scenery along the Dorset coast, most notably around St. Alban's Head and near Lulworth Cove.

Inland, the Portlandian rocks thin out rapidly northwards. From about 240 feet on the Dorset coast they thin to about 30 feet in the neighbourhood of Aylesbury in Buckinghamshire, where they are last seen behind the Bugle Inn at Hartwell.

At the very top of the Jurassic succession in Britain are the **Purbeck Beds**, named after the so-called 'Isle of Purbeck' in south-east Dorset, where it is best developed. It also forms a thin capping on Portland Bill, though much of this has been removed in the quarrying of the building stone beneath. Most of the Purbeck Beds are limestones, laid down in fresh-water lakes and lagoons. This is shown by their abundant fossils, which are mainly molluscs exactly like forms which live in fresh water today. The best known of these limestones is the misnamed 'Purbeck Marble'—actually a limestone almost entirely composed of fresh-water snails. This rock is attractive when polished and has been much used for ornamental work in churches, notably in many of the tombs in Westminster Abbey. Fossil wood is quite abundant here and there, and just up to the east of Lulworth Cove is a 'fossil forest' where the stumps of trees can be seen still in position.[1] Below the trees is a thin 'dirt bed' which is interpreted as the actual soil in which they grew.

The Purbeck is one of the few formations in which vertebrate fossils are fairly common in places. In Durlston Bay—round the headland from Swanage—it is easy to find fish-teeth, scales and turtle bones. An important discovery was made here some years ago, when the foundations were dug for a restaurant on the cliff top. A pocket of small bones was found in a 'dirt bed', which proved to include several primitive mammals (see Figure 41). Jurassic mammals are only known from three localities in the world—the Middle Jurassic at Stonesfield, the topmost Jurassic at Durlston Bay and a small pocket at about the same level in Wyoming. They were insignificant animals, no bigger than cats, but these meek creatures were due to inherit the earth from the giant reptiles by which it was then dominated.

In the middle of the fresh-water deposits there is one good marine layer. This is packed with contorted oyster-shells and is called the 'Cinder Bed' because of its rough, black appearance; it is the last marine episode in the British Jurassic. Like the Portlandian, the Purbeck Beds thin out rapidly northwards; from 400 feet at Durlston Bay, they are down to less than 90 feet in the Vale of Wardour and to only 12 feet near Aylesbury. Eastwards, they are known to thicken under later deposits. They appear again in three small inliers in east Sussex—notably at Battle (of 1066 fame). The continental nature of the beds goes a stage further here and includes evaporites, formed by the evaporation of enclosed lagoons or lakes in a hot climate. Gypsum (calcium sulphate) is mined near Battle for the manufacture of 'plaster of Paris'.

Thus ended the Jurassic record in Britain, consisting as it did of almost unbroken marine sedimentation and without any noteworthy disturbances of any kind.

The same story holds true for most of Europe, except that down in the south, as in the Trias, the shelf-sea conditions pass into those of the 'Tethys' geosyncline. The Jurassic System takes its name from

[1] It is as well to visit this locality on a Sunday, as it forms part of the target area of a tank gunnery range which is in use on other days, but it is hoped that it will shortly be demilitarized.

the Jura Mountains which stretch in a broad banana-shaped outcrop from Savoy in south-east France through western Switzerland into south-west Germany (see Figure 58). These are great rolling well-wooded hills mainly of Jurassic and Cretaceous limestones. They are the outer ramparts of the Alps and mark the margin of the Mesozoic 'Tethys'.

In the Alps the Tethyan Jurassic is confused and altered by later earth-movements. It seems to have functioned as a faunal reservoir from which successive migratory waves populated the shelf-seas. Many of its forms, however, never left the region and others which are common in the shelf-sea deposits are not found here. Among the Alpine brachiopods, for example, are many curious forms, quite unknown elsewhere, and apparently adapted to an unusual environment which it is difficult to interpret.

Jurassic rocks are found all round the Mediterranean. The Rock of Gibraltar is a great chunk of Jurassic limestone. In north-west Italy, not far from Pisa, Lower Jurassic limestones have been metamorphosed to form the pure white Carrara Marble, which is seen all over the world in statues, tombstones and washstands.

The title of this chapter is 'The Long Quiet Episode' and this is very suitable for the Mesozoic era in Europe. It is important to realize, however, that even this, the quietest of episodes, was not the same everywhere. In western North America, for example, the Jurassic was a time of great volcanicity, major earth-movements and the emplacement of huge granite masses. In other words, the European plate was sailing peacefully at this time, whilst western North America was grinding against the Pacific plate.

THE CRETACEOUS PERIOD

In southern England the lakes and lagoons of the Purbeck were quietly followed by the river muds and sands of the early Cretaceous. Three rivers flowed into what is now south-east England—one from the west, one from the north-west and one from the north-east. Each river brought its own distinctive contribution of rock-fragments and minerals from the area of its headwaters. Thus the river from the west brought mineral grains which show that the Dartmoor Granite, after emplacement at great depth in late Palaeozoic times, was at last uncovered by erosion.

166

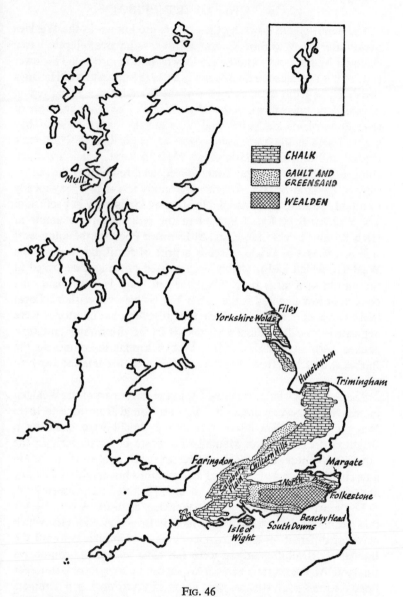

FIG. 46

Map showing the main outcrops of Cretaceous rocks in Britain. The lines in north-east Ireland indicate Cretaceous outcrops which cannot be differentiated on a map of this scale.

The deposits laid down by these rivers are known as the **Wealden Series** after the Weald of Kent and Sussex—the great elliptical area enclosed by the North and South Downs (see Figure 46). The lower beds of the Wealden are mainly sandy and are known as the 'Hastings Beds'. These form the centre part of the Weald, the wooded region of sandstone hills known as the 'High Weald'. The sands are those of river flood-plains and deltas and there are no marine fossils. They are well seen as the upstanding blocks of 'High Rocks', 'Harrison's Rocks' and others near Tunbridge Wells in Kent, which are much climbed by London-bound mountaineers as a relief from the monotony of lowland Britain. At times the sandy rivers gave way locally to muddy swamps in which plants of many kinds grew in profusion. These can still be found upright in the sediment, for example at High Brooms brick-pit in Kent. At Hanover Point on the south-west side of the Isle of Wight, there is a drift of fossil pine-trees in the Wealden, which had probably been carried down and deposited at the mouth of a large river. The 'pine-raft' as it is called, can only be seen at low tide and is the oldest bed visible on the island. Local lakes came and went, in which thin fossiliferous limestones were deposited. So-called 'Sussex Marble' is in fact freshwater limestone packed with snail shells. It is very like similar limestones in the Purbeck, and like them has been much used for interior work in churches.

There are some famous bone-beds in the lower part of the Wealden in which water-worn dinosaur bones are found from time to time. Many of the earliest discoveries in the Weald were made by a Brighton doctor Gideon Mantell who wrote of fossils early in the nineteenth century as 'Medals of Creation'. His published diaries are a fascinating record of the life of an ambitious but embittered doctor and palaeontologist in the days of blood-letting and place-seeking.

The most important of all the Wealden deposits used to be the thin seams of iron-stone and layers of iron-stone nodules which occur at different levels in these continental deposits. Up until the Industrial Revolution, these were the chief source of iron-ore in England. All over certain parts of Weald can be seen the old 'Hammer Ponds' where small streams were dammed up to work trip hammers in tiny primitive ironworks.

It is difficult nowadays to imagine this gloriously peaceful countryside as once England's industrial 'Black Country'. The thick Wealden forests which grew on the clay formations were cut down for charcoal to smelt the iron. Only the discovery and development of coal and iron close together in the Coal Measure took the heavy industries elsewhere at the beginning of the nineteenth century. The last furnace went out at Ashburnham in Sussex in 1828. Nowadays the thin, poor-quality iron of the Wealden is quite valueless from the economic point of view.

Towards the end of the Wealden Epoch, the sands of the Hastings Beds gave way to the **Weald Clay**, a thick, rather monotonous formation which underlies the great horseshoe of low-lying ground between the High Weald and the Downs. The rivers flowing into the area seem to have coalesced into a vast series of lakes and mud-flats. Many geological textbooks talk of a single great 'Wealden Lake', but it was probably never as simple as that.

Northwards the Wealden deposits thin out and disappear against the London ridge—an upland area which stretched eastwards from

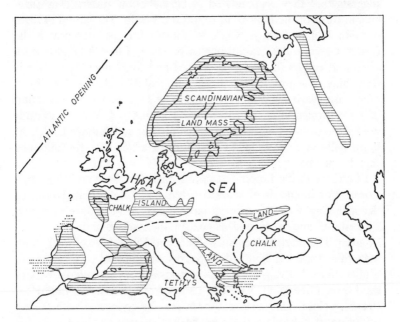

FIG. 47

Reconstruction of the supposed geography of Europe during late Cretaceous times, with the Chalk sea extending from Ireland in the west to Turkey and the Crimea in the east. (Simplified from *A Palaeogeographical Atlas of the British Isles* . . . by L. J. Wills, by kind permission of the author and Messrs. Blackie & Son, Ltd.)

England as far as Germany. Whilst the land deposits of the Wealden were being laid down in southern England, marine sediments were accumulating on the other side of the ridge in the north-east. A northern sea spread into Yorkshire, Lincolnshire and Norfolk with clays, sands and iron-stones, all of which yield ammonites and other marine fossils. At the same time a southern sea—with different faunas—was spreading up from the south. At the very top of the Wealden, a few brackish-water fossils begin to appear, and then with the next division—the **Lower Greensand**—the southern sea spread over the Wealden mud-flats. The two seas, one from the south, one from the north, slowly crept round the west end of the London ridge and eventually met. The deposits of this time are mainly sandy. The 'green' part of the name comes from the green mineral glauconite, which is very abundant in some layers and is a sure sign of their marine origin. The sands are frequently 'false-bedded' which suggests that they accumulated in very shallow water. They form hills in front of the Downs all round the Weald. The largest of them is Leith Hill in Surrey, which is notable for the presence in the Lower Greensand of large masses of **chert**. Chert is almost pure silica and occurs in many sandstones and limestones. Its origin is something of a problem, and the same solution does not apply in every case, but it can often be shown to be a secondary concentration of silica formed after the rocks were consolidated. Such concentrations from percolating solutions often form round suitable nuclei. At Leith Hill the chert seems to have formed around the tiny siliceous spicules of certain sponges.

Elsewhere in the Lower Greensand there are clays and thin limestones. For hundreds of years the thin sandy limestones known as 'Kentish Rag' were quarried at many places between Sevenoaks and Folkestone, and were used as one of the main building stones of London. The original Roman wall around London was largely built with this stone. At Nutfield in Surrey an unusual clay known as 'Fullers Earth' is dug for special purposes from the middle part of the Lower Greensand. Its name comes from its use for 'fulling' or cleansing cloth. This curious deposit may be connected with the resumption of volcanic activity in the British area resulting from the splitting open of the Atlantic.

The Lower Greensand is seen again in the Isle of Wight where it forms the sandstone cliffs around Shanklin and Sandown. It also forms the steep cliffs along the little-visited south-west coast of the island. Thereafter it thins rapidly westwards, as it does northwards against the London ridge. West of London, where the sea broke through between the London ridge and Wales, the Lower Greensand is found in several scattered patches. The most interesting of these is at Faringdon in Berkshire, where the Lower Greensand is a gravelly deposit full of beautifully preserved fossils, especially sponges. This

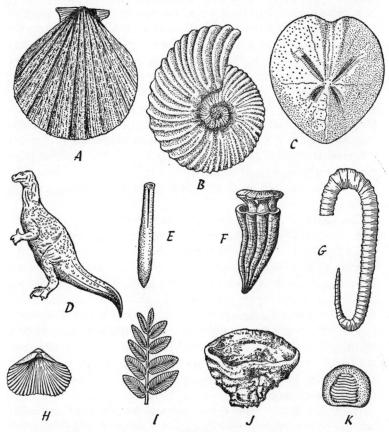

FIG. 48

Cretaceous fossils. *A*: bivalved mollusc—scallop, $\times \frac{1}{2}$; *B*: ammonite (after Institute of Geological Studies), $\times \frac{2}{3}$; *C*: sea-urchin or echinoid, $\times \frac{1}{2}$; *D*: re-construction of a herbivorous dinosaur, $\times \frac{1}{190}$ (the animal was about $20\frac{1}{4}$ feet high); *E*: belemnite guard, $\times \frac{1}{4}$; *F*: aberrant mollusc from the south of Europe, $\times \frac{1}{3}$; *G*: uncoiled ammonite, $\times \frac{1}{2}$; *H*: brachiopod, $\times \frac{2}{3}$; *I*: fern leaf, $\times \frac{1}{2}$; *J*: sponge from Faringdon (after Geological Survey), $\times \frac{1}{2}$; *K*: palate tooth of a shark, $\times \frac{1}{3}$.

171

deposit accumulated in a narrow channel cut in Jurassic clays, and it contains many **derived fossils** from the older strata. These are fossils which have been washed out of their original resting places and included in a later deposit. They can usually be recognized by their worn appearance, but they sometimes cause much confusion.

Above the Lower Greensand come two formations—the **Gault Clay** and the **Upper Greensand**—which illustrate with great clarity how rocks may change their nature laterally from place to place. If one goes to the coast just north of Folkestone harbour in Kent (or looks at it from a departing Channel steamer) one sees a thick clay formation—the Gault—directly overlain by the white Chalk of Dover's famous cliffs. If on the other hand one goes to the Devon coast, near Seaton, one finds a thick sandstone formation under the Chalk—the Upper Greensand—with only a few feet of sandy Gault Clay at its base. The Upper Greensand first becomes recognizable near Sevenoaks in Kent, and from there on westwards it becomes progressively thicker at the expense of the Gault beneath. But these are not really separate formations laid down one after the other; they were contemporary deposits laid down under different conditions.

The Gault is a dark sticky clay, packed with fossils, especially ammonites (see Figure 48). It represents deeper water conditions and is the first formation to pass straight over the ancient rocks of the London ridge without change or hindrance. In deep borings under London, the Gault is always the oldest of the Mesozoic rocks one can be sure of finding; it often rests directly on the Old Red Sandstone. The Upper Greensand on the other hand is a shallow-water deposit with coarse current-bedded sands and thick-shelled fossils. It also contains a great deal of chert. The lateral change in rock type can be attributed to the approach of a shoreline as one moves west. The old rocks of Cornwall probably stood up as an island or as part of a great western continent.

The fossils of the Upper Greensand and those of the Gault differ because of the different conditions under which they accumulated, but there are enough species common to both to show that they were—in a general sense—contemporary. The famous Gault ammonites from Folkestone are mostly preserved in iron pyrites, a mineral which disintegrates under normal atmospheric conditions unless specially treated. Many of the Gault ammonites are odd in appearance, being uncoiled and straight or curiously hooked (see

Figure 48). Since the ammonites were due for extinction at the end of the Cretaceous, these oddities have often been interpreted as the final frantic specializations of a doomed race. Such an association of ideas is both unsound and unsupported by the facts. One cannot literally have racial senility, for every organism is always perfectly adapted to its environment.

Along the Dorset coast, the Gault Clay cuts progressively across older and older formations in the same way as it does under London (see Figure 49). At Swanage it rests on the Lower Greensand, at Lulworth Cove on the Wealden, near Weymouth on the Purbeckian and Portlandian and so on across all members of the Jurassic down to the Lower Lias at Lyme Regis and then on to the Rhaetian and Keuper Marl near Seaton. It is a perfect example of the phenomenon of **overstep**—i.e. the younger formation overstepping the older—and represents a great marine transgression which is known in many parts of the world.

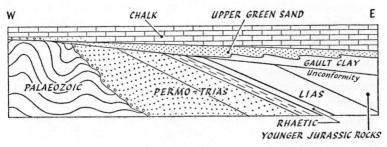

FIG. 49

Diagrammatic representation of the overstep of the Cretaceous rocks on to older formations in south-west England.

For some reason the geologist cannot as yet explain, the seas of the world seem to have overflowed and flooded huge land areas at roughly the same moment at the beginning of Upper Cretaceous times. Thus in the Atlas Mountains of North Africa and the Gulf Coast of North America, we can see the same overstep that we study on the Dorset coast. This is the same transgression that was mentioned in the geological history of Gondwanaland.

At the top of the Cretaceous comes what is probably the strangest and yet most familiar formation in the British record. This is the **Chalk** which forms the off-white cliffs of Dover, the 'Seven Sisters' near Eastbourne (see Plate 15) and the 'blunt, bow-headed, whale-backed Downs' of southern England. The North Downs and the South Downs encircle the Weald and meet in Salisbury Plain to the west (see Figure 46). The Chalk rises again in Portsdown Hill behind Portsmouth and in the sharp backbone of the Isle of Wight from Culver Cliff to the Needles, which continue on into Dorset. A broad sweep of rolling, thin-turfed, streamless Chalk country extends up via the Berkshire Downs and the Chilterns into East Anglia and then on into the Lincolnshire and Yorkshire Wolds. All these hills are formed of the white, soft limestone which we call Chalk. At its base it is grey and muddy as it passes into the strata below, but the upper part is remarkably pure. It is often highly fossiliferous. Ammonites are common in the lower part, but the most useful fossils of the rest of the Chalk are sea-urchins (see Figure 48).

A Margate doctor—A. W. Rowe—made one of the finest studies of any formation, near the beginning of the present century. He described all the white Chalk cliffs around England in great detail and also produced a classic of evolutionary research in a study of one of the Chalk sea-urchins.

The Chalk has long been a problem because its grains are too small to study even under the most powerful optical microscope. Tiny fossil foraminifera are locally abundant and these are like modern forms which float near the surfaces of the oceans. The skeletons of the latter accumulate in vast numbers on the deeper parts of the ocean floor, and for this reason many workers have thought the Chalk to be a deep-sea deposit, but this does not fit in with the evidence from other fossils which suggests a depth of no more than about 100 fathoms. Since the last war, study of the Chalk with an electron microscope has shown it to be made up almost entirely of the hard parts of minute, very simple, floating plants known as **coccoliths**.

The problem of the Chalk today is not so much where the material came from, as how other material was kept out. The remarkably pure organic Chalk is almost completely without any trace of land-derived sediment. The land surrounding the Chalk must have been very low-lying and not undergoing erosion. It may have been a hot desert, and reddish material found at the base of the Chalk at various places supports this idea, as does the finding of desert-type sand-grains in the Chalk on the island of Mull. If no appreciable amount of material was coming from the land, then inevitably, though very

very slowly, the hard parts of the animals and plants living in the sea would build up a deposit on its floor. The Chalk is over 1,500 feet thick in places in England, and this gives some idea of the length of time it must have taken to form.

The old three-fold subdivision of the Chalk was into the 'Grey Chalk' at the base, the 'White Chalk with few flints' in the middle and the 'White Chalk with many flints' at the top. This does not always hold true, but is a useful guide. The black **flints** of the upper part occur as continuous bands and as lines of nodules. They are very much harder than the Chalk itself and survive after erosion to provide the vast bulk of the pebble beaches of the south coast (see Plate 15). Flint pebbles from the Chalk are also abundant in Tertiary and Pleistocene deposits. Flint is pure silica and a variety of chert, which has already been discussed. It can be distinguished from the chert which occurs in other formations by its smoother, blacker surface and the clean shell-like fractures it shows when broken.

At certain levels in the Chalk there are hard nodular bands with much green glauconite and distinctive fossils. These bands indicate that the floor of the Chalk sea was uplifted far enough at times for the sea-floor sediments to be affected by wave action. This is further evidence that the sea was never very deep.

The Chalk sea extended farther than any previous sea in the Mesozoic Era. In Britain, all the barriers were swamped and the sea spread as far as the Hebrides and Northern Ireland. Some geologists think that the Chalk once extended right across the ancient rocks of Wales submerging them for the first time since the Silurian. This is suggested, not by actual deposits, but by the fact that the mountain-tops are planed off at a uniform level as though by the action of a past sea.

The same sea stretched right across northern Europe to southern Russia and Turkey (see Figure 47). It appears to have been particularly deep in the area of Poland and the deposits there are exceptionally thick. A similar deposit of Upper Cretaceous age is found in the southern part of the U.S.A.

All this time, away in the south, the 'Tethys' still persisted, all the way from the West Indies to south-east Asia. In this sea the ammonites survived right until the end of the Cretaceous and were then suddenly extinguished together with several other groups. Also in the 'Tethys' there lived a strange group of aberrant bivalved molluscs, often huge in size and modified to look more like corals than molluscs (see Figure 48).

175

THE LONG QUIET EPISODE

The most important change which took place in the life of the world during Cretaceous times was the sudden springing into dominance of the flowering plants. All through the Mesozoic so far, plant life had consisted mainly of conifers and their allies, cycad-like plants and ferns. Suddenly, at about the beginning of Upper Cretaceous times, the flowering plants appeared and spread rapidly. They soon outnumbered all the others, as they do today. There are reports of flowering plants earlier in the record, but most of these have been disproved. Flowers, and the resultant completely enclosed seeds, are certainly the most highly evolved and the most efficient method of plant reproduction, but they are unfortunately extremely poor material for fossils.

The most common plant fossils are leaves, and these are often a very poor guide to the nature of the plant. It is a similar problem to distinguishing the earliest mammals from the mammal-like reptiles. Nevertheless, we can be fairly sure that common among the earliest flowering plants were forms such as the Magnolia which are generally accepted as simple, primitive types. But before the end of the Cretaceous Period, more advanced families, including the all-important grasses, had already evolved.

In the animal kingdom, the reptiles continued to dominate the world up until the end of the Mesozoic. In Britain, many dinosaur remains are known from the earlier Cretaceous strata in the Weald and the Isle of Wight. In Belgium, during the excavation of a mine shaft, the skeletons of no fewer than twenty-nine giant Cretaceous herbivores were found tumbled together. Apparently their bodies had been washed into a deep ravine cut through the coal-bearing Carboniferous rocks. In western North America, many huge Cretaceous dinosaurs have been found, including the largest of all carnivores and some curious ostrich-like forms.

A few mammal teeth are known from the Lower Cretaceous—first found in the English Wealden by Charles Dawson who also allegedly found the notorious Piltdown skull. More mammals are known from the Upper Cretaceous, from Wyoming and Mongolia. These are all small forms and include both marsupial or pouched mammals and the more advanced placental mammals. The fascinating problem of how these humble creatures inherited the earth from the giant reptiles, and why so many groups became extinct at the end of the Cretaceous, will be discussed in the next chapter.

176

IX

THE BEGINNING OF MODERN TIMES

This chapter deals with the Palaeogene Period which began the Caenozoic Era and included the Eocene and Oligocene Epochs. It was that part of the Tertiary which preceded the main movements of the Alpine Orogeny.

THE SEDIMENTS

Not far west of London, near the Middlesex/Buckinghamshire border, is the village of Harefield. Nearly opposite the church there is a large old quarry now, regrettably, being used as a rubbish-dump. This is one of a series of workings where the Chalk was quarried and made into lime for cement. As such they are like hundreds of other quarries in south-east England. On top of the Chalk at Harefield, however, there is a succession of completely different sediments—gravels, sands and clays—which the average person would not regard as 'rocks' at all.

If one stands well back from the face of the quarry and studies the bands of flints which follow the bedding planes in the Chalk, one can see that they are cut out one by one at the junction between the hard Chalk and the softer sediments on top. In other words, the junction is an unconformity and this particular unconformity is one of the most important breaks in our record, for it marks the end of the Mesozoic Era and the beginning of the Tertiary.

At the end of the Mesozoic there were major changes in the geography of many parts of the world. There was an orogeny going on at this time in the western part of North America and the Rocky Mountains were being uplifted. In northern Europe the floor of the sea was raised up and the Chalk was subjected to erosion. In Britain, the top of the Chalk is lost because of this erosion. The youngest

Chalk left is on the coast at Trimingham in Norfolk, though it has been difficult to study because of a minefield left over from the last war and tumbling clay cliffs above.

Elsewhere, notably in Denmark, much later beds of Chalk are preserved and pass up almost imperceptibly into the Palaeogene above. In the south of Europe—in Italy for example—the 'Tethys' sea continued as before and geosynclinal sedimentation went on without break until much later.

The great problem of the Mesozoic/Caenozoic junction is why at this moment in time, many major groups of animals should have suddenly suffered eclipse or extinction. Obvious examples are the dinosaurs, the marine reptiles and the ammonites, all of which disappeared abruptly from the record. Others—for example the brachiopods—became relatively insignificant. In Britain and in other places where there is an obvious gap in the record at this point, the rocks change completely in character and the faunal change is not so difficult to accept. In fact we may argue that the change only seems to be abrupt because of the gap in the stratigraphical record. But the change was in fact universal and no easy explanation seems to be sufficient.

Certainly we cannot postulate a spectacular catastrophe which wiped out vast populations, for there is no suggestion of this in the rocks and plenty of animals and plants survived the climax unaffected. The most popular suggestion is that certain animal groups could not adapt themselves to the great changes in geography and environment which occurred at this time. One hears of increasingly arid conditions causing the drying-up of the wet swamps favoured by the giant herbivorous dinosaurs. These creatures, with many millions of years of specialization behind them, could not live anywhere else. As they died out, so also perished the carnivorous dinosaurs who depended on them for food. In North America, where there are numerous dinosaurs in late Cretaceous rocks, such climatic changes can be correlated with earth movements in the Rockies. No orogeny appears to have been world-wide in its effects, however, and it is difficult to believe that this explanation is sufficient.

Disease has been suggested as a possible cause of the widespread extinctions, but as we have clearly seen with myxomatosis, it is extremely unlikely that an epidemic could wipe out a single species, let alone a major group. Another very important and widespread event which may have brought about the downfall of the dinosaurs

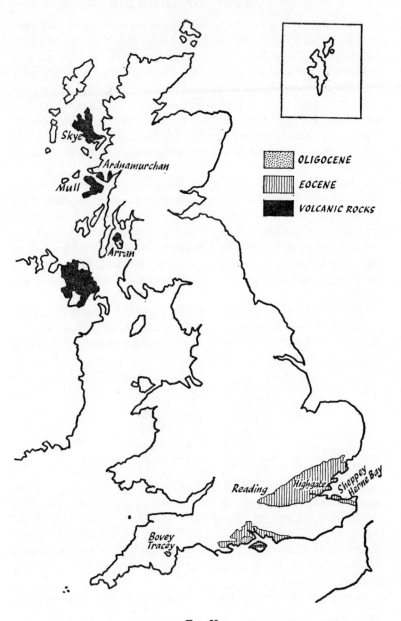

OLIGOCENE

EOCENE

VOLCANIC ROCKS

Skye

Ardnamurchan

Mull

Arran

Reading

Highgate

Sheppey

Herne Bay

Bovey Tracey

FIG. 50
Map showing the main outcrops of Palaeogene rocks in Britain

THE BEGINNING OF MODERN TIMES

was the great change in the plants which occurred early in Upper Cretaceous times. Since the herbivorous dinosaurs must have been adapted to a diet of ferns, conifers and cycad-like plants, the major replacement of these by the flowering plants must have considerably upset the reptiles' feeding habits and perhaps indirectly caused their extinction. This, of course, only pushes the problem farther back in time, and one must then still find a cause for the plant changes. Whatever happened, the more adaptable mammals were able to ride out the geological storm and to take advantage of the clear water ahead.

There still remains the problem of the changes in the marine faunas which also occurred at this time. These can hardly be blamed on increasing aridity or changes in the land flora, but they are just as spectacular as the changes in the land faunas and much more important to the ordinary working geologist. With the reptiles it can be argued that they were replaced by the more efficient mammals, both on land and in the sea, but no obvious successors replaced the ammonites.

The post-Mesozoic part of the stratigraphical column is usually subdivided into five or six so-called systems. Their names were intended to refer to the relative abundance in them of species still living today. They are as follows (oldest first):

Eocene: 'dawn recent'[1]
Oligocene: 'few recent'
Miocene: 'less recent'
Pliocene: 'more recent'
Pleistocene: 'most recent'
Holocene: 'all recent'

All these are tiny divisions compared with the systems of earlier times and they cannot strictly be regarded as such. The most satisfactory arrangement in the author's opinion is to group them in three pairs of 'Series'—the Lower Tertiary or **Palaeogene System** (Eocene and Oligocene), the Upper Tertiary or **Neogene System** (Miocene and Pliocene) and the **Quaternary System** (Pleistocene and Holocene).

As one goes up this later part of the record, there is certainly a

[1] In some parts of the world, notably North America, a further division—the **Palaeocene**—is recognized before the Eocene; also many geologists do not use the term 'Holocene'.

180

rapid increase in forms closely resembling those of the present day, but one man's species is another man's genus and we cannot now be as dogmatic about proportions as were the earlier geologists.

The Palaeogene sediments of the British Isles are confined almost entirely to two areas—the London Basin which extends down the Thames Valley, and the Hampshire Basin which includes the northern half of the Isle of Wight (see Figure 50). These two basins were formerly linked with one another and with other basins in France and Belgium.

The sea which spread across southern England after the uplift of the Chalk came from the east. The sea which invaded France and Belgium at the same time came from the west. It seems that throughout Palaeogene times there was a sea in north-west Europe, centred in the North Sea, which periodically expanded and retreated its margins (see Figure 51). As a result we have in southern England (and in France and Belgium) a series of 'cycles of sedimentation' each of which records a single coming and going of the sea.

What is probably the finest and most complete section in Palaeogene rocks in Europe is in Whitecliff Bay, at the east end of the Isle of Wight, just south of Bembridge. At the south end of the bay is the sheer white mass of Chalk that is Culver Cliff. If one goes to the farthest possible point at low tide one can see the lines of black flints along the bedding planes in the Chalk rising almost vertically. The softer Eocene rocks which form the back of the bay are similarly dipping at a very steep angle, whilst the Oligocene rocks at the north end of the bay flatten out in the cliff before Bembridge Harbour. All these sediments were turned up on edge during the mid-Tertiary earth-movements which formed the Alps. This means that one crosses the entire Eocene succession and a great part of the Oligocene as well, in a few hundred yards.

Within the Eocene sediments of Whitecliff Bay one can see five distinct marine horizons, usually clays, packed with fossils of many kinds. Above the marine clays there are often curious laminated beds. These are thin alternations of clay and sand with great quantities of drifted plant material, and are probably the deposits of tidal mud-flats like those around the Wash at the present day. They are followed in turn by variously-coloured pale sands which were probably formed on dry land. Ideally, these are succeeded by laminated beds again and then by marine clays completing the cycle.

The Oligocene sediments which follow were nearly all laid down in

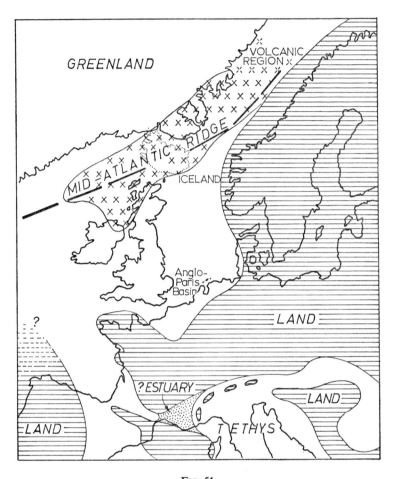

FIG. 51

Reconstruction of the supposed geography of north-west Europe during Palaeo-gene times (after Wills and others).

fresh or brackish water, though there are three marine horizons. They include the Bembridge Limestone which stands out clearly in the cliff and forms the natural promenades at the north end of White-cliff Bay. The limestone is packed with freshwater snails and is used locally for building purposes. Oligocene clays, marls and thin lime-stones underlie all the gently undulating northern half of the Isle of Wight. They also extend on to the Hampshire mainland around

182

Lymington. None are known in the London Basin or elsewhere in Britain, except for the unusual deposits at Bovey Tracey in Devon, which will be discussed later.

Returning to the Eocene sediments of Whitecliff Bay, these change as they are traced westwards, and by the time they reach Alum Bay at the west end of the island, they are quite different in appearance. It is very instructive to compare the two bays, which are both de-limited to the south by the Chalk backbone of the island and to the north by the more gently dipping Oligocene strata. At Whitecliff Bay there are five marine horizons in the Eocene, at Alum Bay there are only two. Instead, at the latter, there are much greater thicknesses of land deposits and the missing marine phases are only represented, at most, by the laminated tidal flat type of deposits. The continental sands of Alum Bay, which are highly coloured due to iron-staining, are one geological phenomenon which is known to every visitor. Thousands of summer holiday-makers buy excessively expensive test-tubes, glass pens or lighthouses to fill with vari-coloured sands from the tumbling cliffs, or if exhausted with the manifold delights of the island, buy them already filled.

Besides the famous coloured sands, there are many other interest-ing features in the Eocene continental sediments. There are beds of large flint pebbles produced when there was a speeding-up of the erosion of the newly-emerged Chalk hills. There are beds of lignite or fossil wood, even one in Whitecliff Bay which approaches the quality of a poor coal and which is underlain by a rootlet bed. The best-known feature of Alum Bay to geologists is the presence of **pipe-clays** with well-preserved leaves. These are seams—often almost white—of remarkably pure clay which was much used in the past for making clay pipes. Similar clays were formerly worked on the main-land for this purpose—for example at the beauty spot known as the 'Blue Pool' near Corfe Castle in Dorset. Leaves can be found quite easily in one of the pipe-clays in Alum Bay, and though difficult to identify with certainty, closely resemble the leaves of plants which live today in tropical and subtropical climates.

All these features clearly show that the Eocene sediments in Alum Bay were laid down under more continental conditions than those of Whitecliff Bay. In other words, the farther we go to the west, the more evidence we have of dry-land conditions and the less we have of marine conditions. This is still more marked on the Hampshire mainland around Bournemouth and Boscombe, where the continental sands form the postcard scenery of the deep valleys or 'chines'. Farther west still, the Eocene is only represented by gravels. The sea therefore clearly came from the east and some invasions were more extensive than others.

The same phenomenon can be demonstrated in the Eocene beds of the London Basin, where the process started even earlier. The earliest Caenozoic marine incursion laid down fossiliferous sands and clays in East Kent (for example at Herne Bay). These strata can be traced into east London, where the sands proved suitable for moulding purposes and were responsible for the siting of Woolwich Arsenal. There were many sections like that at Erith, shown in Plate 16. The upper part of this basal Eocene formation then develops a brackish

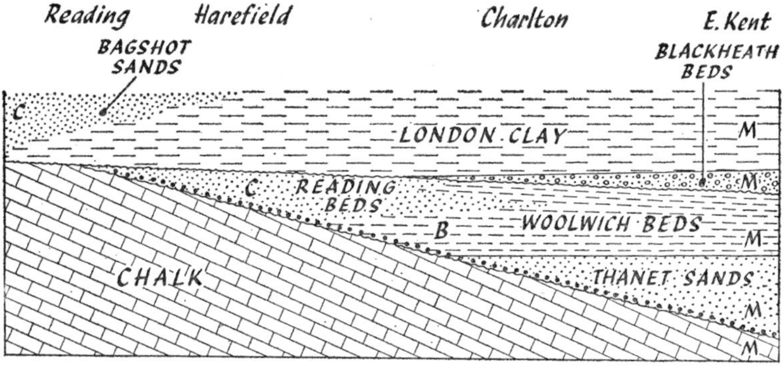

Fig. 52

Diagram to show change of facies of Eocene sediments in the London basin. M = marine.　　B = brackish.　　C = continental.

water fauna. This was formerly well seen around the old Charlton Chalk quarries, where the Charlton Athletic football ground now stands. Only one section now remains and in this, the beds abound in the sorts of shells which live today in estuaries and similar environments. Farther east still, stretching to Reading and beyond, are sands and mottled clays which were clearly laid down by a river system coming from the west. The only fossils in these beds are occasional leaves and other plant fragments. Thus strata of the same age differ markedly from place to place even within a short distance, and so do the fossils which they contain. The way in which these early Eocene beds change laterally in the London Basin is shown diagrammatically in Figure 52. These changes make Palaeogene stratigraphy difficult but at the same time particularly interesting.

The first marine invasion of the Hampshire Basin, and the main one of the London Basin, produced the thick marine formation known as the **London Clay**. This underlies the greater part of the capital and, since it is a thick clay formation, proved very convenient for the digging of London's underground railways. It is, however, difficult to find an exposure of this ubiquitous clay except in the deeper temporary excavations which go below the sands and gravels of the River Thames. Westwards in the London Basin, the London Clay (like the earlier formations) tends to change its nature and the upper part at least passes into a continental deposit known as the **Bagshot Sands**. These underlie the greater part of the west end of the London Basin and give rise to the rhododendron and birch-tree country around Bagshot, Sandhurst and Farnborough. In this case, as in many others, the vegetation gives a clear indication of the nature of the rocks beneath.

Probably the best place for studying the London Clay is on the so-called 'Isle' of Sheppey in the Thames estuary. Here, mixed up with the usual marine fossils, there occur the remains of river-dwelling animals such as crocodiles and an almost incredible abundance of twigs, nuts and fruits. These must have been washed down by a river to settle on the sea-floor far off-shore. It is significant that there are no leaves or other perishable parts which would not survive such a journey. The fruits have been studied in great detail and show a remarkable resemblance to floras living today around the seaboards of the Indo-Malayan region.

As one goes up (in a figurative manner) through the Palaeogene and Neogene rocks of north-west Europe, there are fewer and fewer tropical and subtropical plants. This is one of several lines of evidence which show that there was a progressive deterioration of climate throughout this long period of time.

The London Clay is represented in Belgium by the Ypres Clay, so intimately and tragically known by our soldiers in the First World War. It is also known in Denmark, where it contains numerous bands of volcanic ash. This is an indication of more violent events which were going on in the North Atlantic region at this time; these will be discussed later. No marine equivalent of the London Clay is, however, known for certain in the Paris Basin. It seems probable that the London Clay sea spread into England from the north—from the area of Denmark and north-west Germany.

The later Eocene seas on the other hand flooded right across the Paris Basin and connected with the 'Tethys' geosyncline which still persisted away to the south. This is shown by the presence in the upper Eocene clays of the foraminifera known as **Nummulites** (see Figure 55). These are exceptionally large for the group and they evolved very rapidly. One of the largest species—about the size and shape of a ½p coin—is very common in Whitecliff Bay, as are several other species. They were most abundant in the 'Tethys' where they formed the main constituents of thick Eocene limestones, for example in Egypt, where 'nummulitic limestone' was used to build the Great Pyramid of 'Cheops'. Besides the nummulites, there is a tremendous abundance of fossils—especially 'sea-shells'—many of which show strong southern affinities.

Perhaps the most fossiliferous locality in Britain is at Barton, about nine miles east of Bournemouth, where the constantly tumbling cliffs are formed of clays at the very top of the Eocene, full of an amazing variety of molluscs. These belong almost entirely to two classes—the gastropods and the lamellibranchs—which in the Eocene really came into their own for the first time. All through the Mesozoic, the third major group of molluscs—the cephalopods—had dominated the scene with the spectacularly rapid evolution of its various members. During this time the gastropods and lamellibranchs, like the House of Lords 'did nothing in particular and did it very well'. They were locally abundant, for example, in some of the Jurassic shell-banks of oysters, but never played a major role in the record. Their place in the modern world was then taken by the brachiopods, which abounded in the shallows of the Palaeozoic and Mesozoic seas. At the beginning of the Palaeogene, the superior feeding system of the bottom-living molluscs at last got the upper hand in the battle for survival and the brachiopods slipped into their present insignificant position.

The Eocene beds of the Paris Basin are probably even more fossiliferous than those of England and were the medium for some of the earliest scientific palaeontology. The continental beds in the 'cycles of sedimentation' here contain evaporite deposits produced by the evaporation of enclosed bodies of water under a hot sun. 'Plaster of Paris' was made from gypsum in Eocene beds that were once quarried around Montmartre. These also yielded the bones of many early mammals which were described by the great French anatomist Baron Cuvier, who has been called the founder of vertebrate palaeontology.

All through the Palaeogene and Neogene Periods, mammals were evolving very rapidly on the land surfaces and filling all the gaps left by the extinct reptiles. They went even further in response to the changes in vegetation. One of the most important steps was the evolution of grazing animals such as the horse family, that is, forms with teeth adapted to feeding on the newly-evolved grass. These contrast markedly with the simpler browsing animals which could only feed on soft leaves and shoots. Meanwhile, other groups of mammals were evolving in other directions. Some produced the long line of flesh-eating carnivores, some produced the flying bats, some went back to a marine environment and produced the almost incredibly specialized whales. One tiny group went on in undistinguished isolation to produce the anthropoid apes and man.

It has already been said that Oligocene deposits in Britain are almost entirely restricted to the north side of the Isle of Wight and adjacent parts of the mainland. The only other deposit of this age in England is the small Bovey Tracey basin on the east side of the Dartmoor Granite in Devonshire. This deposit was well seen in a large pit opposite the racecourse at Newton Abbot where there are thick white clays produced by the disintegration of the neighbouring granite and thick black lignite produced by the accumulation of plant debris. There are no fossil animals in the deposit, but the plants suggest a very late Oligocene age and this is in fact the nearest approach we have to a Miocene deposit in Britain.

VOLCANIC ROCKS

An entirely different kind of record of Palaeogene times is found in what is called the 'Tertiary Volcanic Province' (see Figure 50). Over a very large area around the eastern North Atlantic there were vast outpourings of basalt—black basic lava—and associated volcanic rocks. They are found today over a large area of north-east Ireland, in Mull, Skye and many smaller islands of the Inner Hebrides, on the adjacent mainland peninsula of Ardnamurchan, in Iceland, Spitzbergen, the Faroes and Greenland. The associated intrusive rocks are found over an even wider area.

This was the first volcanicity in the British area since the Palaeozoic. All through the Mesozoic era there had been quiet sedimentation without so much as a hint of a lava flow or an ash band; then suddenly, at the beginning of the Caenozoic, there was this great outburst of activity. In Britain, it did not last long; the volcanoes were extinct before the end of the Palaeogene, but in Iceland it went on all through the Neogene and is, in fact, still going on at the present day.

It is difficult to think of the hard igneous rocks of the wild mountain scenery in the volcanic districts of Britain as being the contemporaries of the soft, low-lying, urbanized sediments of the London and Hampshire Basins. Yet there is no doubt that they accumulated at the same time. In the Hebrides and elsewhere there are patches of fossiliferous Mesozoic rocks preserved under and by the lava flows. There are even fragments of them caught up in the vents of the old volcanoes. Also, here and there between the lava flows, are thin beds with plant remains of Palaeogene age.

The volcanic activity took several forms. Firstly there were the outpourings of basalt—flow upon flow of lava—building up thousands of feet of hard, black rock. These are seen particularly in Mull, northern Skye and north-east Ireland. More than 6,000 feet of basalt still remains in places on Mull in spite of millions of years of erosion. The lava seems to have been poured out on a land surface, not under the sea, and the top of each flow is often weathered and reddened by the effects of the contemporary climate. In western Mull a geologist found the charred remains of a tree, 40 feet high, standing upright in the basalt; it must have been engulfed and killed by the molten lava.

The plant beds mentioned above occur on the weathered tops of lava flows and sometimes even approach the consistency of very poor coals. The most famous ones are on Mull where the plants are chiefly represented by leaves and pollen grains. These suggest a warm climate, but there has been some dispute about their exact age. The snag is that they have to be compared with floras at other latitudes, for example in Greenland and southern England, and there is every probability that the floras migrated southwards with time due to a deteriorating climate. Plant evolution only went on very slowly and the fragments found in Palaeogene deposits are very like familiar living genera such as the hazel and the oak.

The lavas probably flowed out of huge central volcanoes which have since been obliterated, but it has also been suggested that they may have come from long, narrow fissures in the same way as some do in Iceland. The largest single area of Palaeogene basalt in the

British Isles is in north Ulster. Here there is the famous locality of the 'Giant's Causeway' near Portrush, where the basalt in cooling has produced a geometrically perfect staircase of regular hexagonal columns. Similar columns are seen forming the sides of Mendelssohn's 'Fingal's Cave' on Staffa in the Inner Hebrides. In Northern Ireland, just below the level of the 'Giant's Causeway' basalt, there is a reddish deposit between the lava flows which includes a thick bed of **bauxite**. This is a valuable source of aluminium and iron and was probably produced by the weathering of the basalt under humid tropical conditions.

In all the volcanic areas, the successive lava flows weather out like steps on the sides of the hills; this is known as **trap topography**, 'trap' being the old name for a solidified lava. In Mull there are local pillow lavas, and these are thought to have formed where the lava flowed into lakes lying in a huge old extinct crater. Generally speaking, however, there is not much evidence of the contemporary craters or fissures from which the lavas were erupted.

The basalts were followed by the intrusion of massive igneous bodies, both acid and basic, which may be the deeply eroded foundations of old volcanoes. These form mountainous areas such as the Mourne Mountains and Slieve Gullion in Ireland, the central mountains of Mull and the Cuillins in Skye. Each has its own complicated history. Thus in Mull, the centre of igneous intrusion moved progressively with time towards the north-west. On the peninsula of Ardnamurchan, the centre of activity shifted westwards and then east again.

In Skye there were only two main centres of intrusion—one basic and the other acid. First there was the massive intrusion of the basic rocks which form the black Cuillin Hills, probably the wildest and most rugged scenery in the British Isles. Then there came a series of acid intrusions—of granitic rocks—which form the more rounded 'Red Hills' to the east.

Even more interesting than the major intrusions are the associated minor intrusions, that is, sills and dykes of various kinds. Two types of these which are very obvious on a geological map are **ring-dykes** and **cone-sheets**, which show up as clear circular structures around the central intrusions. Ring-dykes are thin, circular, inclined sheets, sloping steeply outwards (see Figure 53). They are thought to have been produced when great masses of the crust slipped downwards into a magma chamber, allowing molten magma to well up round the edges. This would be a 'cauldron subsidence' such as has already been described in the Devonian rocks of Glencoe and Ben Nevis (see Chapter V). Because of the much greater age of these latter, we see them eroded to a much greater depth. The hollows produced in the land surface were filled with lavas like those of central Mull.

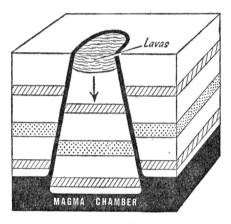

FIG. 53
Highly diagrammatic reconstruction of the supposed form of a ring-dyke.

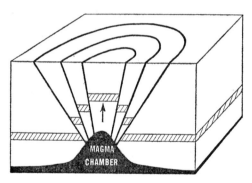

FIG. 54
Highly diagrammatic reconstruction of the supposed form of cone-sheets.

Cone-sheets on the other hand converge downwards (see Figure 54) supposedly to the top of the magma chamber. They are well seen in Ardnamurchan, where there are many of them, one within the other, the inner ones inclined more steeply than the outer. From

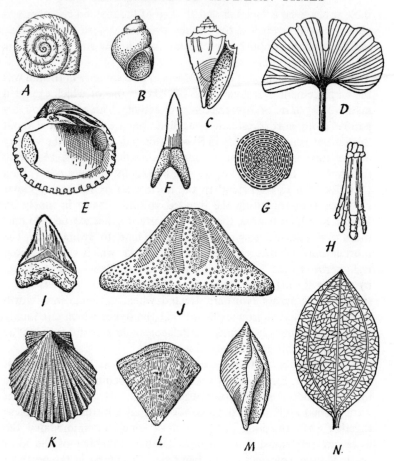

FIG. 55

Tertiary fossils. *A–H*: Palaeogene fossils. *A* and *B*: fresh-water snails (after Institute of Geological Sciences), both × ⅔; *C*: marine snail or gastropod, × ⅓; *D*: leaf of the maiden-hair tree—*Ginkgo*, × ½; *E*: bivalved marine mollusc or lamellibranch, × ¼; *F*: shark's tooth, × 1⅓; *G*: *Nummulites*, a large coin-like foraminifer, × 1; *H*: fore-foot of an early four-toed horse, × ⅕. *I–N*: Neogene fossils. *I*: shark's tooth, × ¼; *J*: large sea-urchin or echinoid, × ⅔; *K*: scallop (after Institute of Geological Sciences), × ⅔; *L*: fan-shaped coral (after Institute of Geological Sciences), × ⅔; *M*: thick-shelled, reef-dwelling gastropod, × ¾; *N*: leaf from the Oeningen beds of Switzerland (after Heer), × ½.

these it has been calculated that the top of the magma chamber was about three miles below the surface when they were formed. They must have been produced by pressure from below and are therefore the exact opposite of ring-dykes.

Probably the last event of the Palaeogene volcanicity in Britain was the emplacement of a great series of dykes, some of which extended many scores of miles from the centre of activity. These are all roughly parallel along north-west/south-east lines, but are concentrated about the various centres, notably in Skye, Mull and Arran. In the vicinity of the great intrusions they occur in fantastic numbers. In Arran, for example, there are 525 dykes in a horizontal distance of less than 15 miles. The dykes extend to the north-west as far as the Outer Hebrides (not forgetting the isolated volcanic centre in lonely St. Kilda). South-eastwards, the more important dykes extend into northern and midland England where they tend to swing west. The most famous of these is the Cleveland Dyke which can be traced right across to the Yorkshire coast and clearly demonstrates its age by cutting through and baking Mesozoic sediments. There is also a set of dykes across Northern Ireland which continue into North Wales, and there is now evidence that the dykes which cut Lundy Island in the Bristol Channel are Palaeogene in age and perhaps the main bulk of this granitic island as well.

The reason for all these volcanic happenings is not now difficult to understand, for we know that the main opening of the Atlantic must have occurred in Tertiary times. Great quantities of sediment were pushed aside to make room for the major intrusions and these together with the long parallel cracks which accommodated the dyke 'swarms' must imply a considerable stretching of the local crust. Similar volcanics are known from Greenland in the north to Portugal in the south. They are all of early Tertiary age (indeed some of them may go back to the Cretaceous) and therefore considerably older than the extinct volcanoes of Iceland and the Faroes, which in turn are older than those of active split in the centre where new ocean floor is being formed at this moment.

X

THE ALPINE STORM AND AFTER

In this chapter there is an account of the Alpine mountain-building movements which occurred during the Caenozoic Era, and of their after-effects in the history of Neogene times.

THE ALPINE FOLDING

The upper part of the Tertiary is most conveniently thought of as one period—the Neogene—though most textbooks still regard it as two—the Miocene below and the Pliocene above. The Miocene is the one part of the stratigraphical column of which we have no sedimentary record at all in Britain. This does not mean that nothing was going on, but that it was a period of erosion rather than deposition. The erosion resulted from the general uplift of the British Isles when the Tertiary crustal disturbances reached their climax in the Alps.

We know now that mountain-building movements or orogenies are not sudden paroxysmal things, but are more like thunderstorms which build up to a maximum and then die away again. Thus the first movements of the Alpine Orogeny can be traced back to the Cretaceous and the last movements can be seen to have lingered on into the Pleistocene. There may even be mountain-building movements still going on in certain parts of the world at the present day, but this possibility will be discussed later.

The Alps and their associated ranges are the most impressive mountains in Europe; the Himalayas are even more impressive in Asia. But this does not mean that the earth-movements which produced them were more severe than any others in the earth's history, it only means that they happened more recently. In North America, the Rockies in the west far exceed the Appalachians in the east in size and grandeur, but this is only because they were formed much later and so have not suffered so much erosion. In Britain, the Caledonian mountains which once stretched up through Wales and across Scotland, and the Hercynian mountains across Devon and Cornwall, were probably like the Alps in their day. These old chains have long ago been worn away. The mountains in the same regions today result from uplift in Caenozoic times and the resistance of the old hard rocks to erosion. The main geological difference between North Wales and East Anglia is that in the former the Lower Palaeozoic rocks have been uplifted and worn into rugged mountains, whereas in the latter they have subsided and been buried beneath hundreds of feet of later sediments.

In North America, a major series of earth-movements—the **Laramide Orogeny**—occurred at the end of the Cretaceous, especially in the Rocky Mountain area. This produced a very clear break between the Mesozoic and the Caenozoic. Similar movements, but on a much smaller scale, occurred in Europe. The main movements here seem to have started early in Palaeogene times and to have built up in a crescendo during the Oligocene and Miocene, to be followed by the diminuendo of the Pliocene and Pleistocene. This can be proved by putting together items of evidence from a hundred different places. Thus in the northern Appenines of Italy, slow thin deposition went on without break from Cretaceous times all through the Eocene and well into the Oligocene. This is shown clearly by the rare fossils. Then, suddenly, the fine-grained, probably deep-water sediments were followed by a coarse detrital bed full of large broken foraminifera of Upper Oligocene age. From then on, sedimentation was coarse and rapid. The whole of the Cretaceous, the Eocene and the Lower Oligocene around Florence amounts in places to little more than 150 feet of sediment, whereas the formation which follows (laid down in just the Upper Oligocene and Lower Miocene) alone may be more than 9,000 feet thick. Clearly something fairly violent must have happened to produce such a sudden change in the amount and coarseness of the sediments being carried down into the sea.

In other areas there is evidence of repeated movements. In the Jura mountains of south-east France, marine Miocene sediments can be seen in places lying horizontally on folded Mesozoic strata. This implies that the major movements were pre-Miocene in age. Elsewhere in the same area there are other patches of Miocene sediment which have themselves been turned up almost vertically by earth-movements which must have happened after their deposition.

194

THE ALPINE STORM AND AFTER

To European geologists the Alpine Orogeny means, above every-thing else, the Alps, and the study of those wonderful mountains has become almost a separate science. For many years now, the geolo-gists of Switzerland and Austria, together with a smaller but distin-guished contingent from adjacent countries, have been working out the fantastically complicated Alpine structures. Their conclusions vary considerably in the degree of complexity recognized, but there is no doubt that we have nothing like it in Britain above the Pre-cambrian. The rocks involved in the Alpine folding are mainly the Mesozoic limestones and shales of the 'Tethys' geosyncline, but great thicknesses of late Palaeozoic strata are also involved. The various sediments are spread out in great piles of recumbent folds from the French Mediterranean coast to Vienna. The intensity of the folding produced considerable metamorphism and erosion has now cut down deeply through the piles of folds to expose the metamor-phosed roots.

The Alpine folds are far more considerable than any we have in England, though the Dalradian rocks of Scotland probably have comparable structures. The anticlinal folds have been pushed right over to become recumbent (see Figure 21), and then the inverted limbs have been so much stretched that they have fractured and the folds have slid along great distances on thrust planes (see Figure 56). Such vast combinations of folding and thrusting are called **nappes** by the Alpine geologists (see Figure 56A). This means literally 'sheets' and refers to the sheet-like nature of the great rock masses which have usually been cut off by erosion from their roots and lie on top of completely different rocks, often much younger in age.

Sometimes the greater part of a nappe has been removed by erosion and only isolated patches remain far from their original position. Such isolated fragments are known as **klippen** and there were many of them in the Alps to puzzle the early geologists (see Figure 56B). The best-known example of a klippe is the dark marble pyramid at the top of the Matterhorn, which rests on an earlier nappe of younger rocks.

All the nappes of the main part of the Alps seem to have come from the south. The study of the different types of sediment among the Mesozoic rocks has helped considerably in unravelling the struc-ture. Thus in places, sediments which were probably deposited on the south side of the Alpine geosyncline seem to have been pushed right over to the north side of the Alps. The Northern Calcareous Alps,

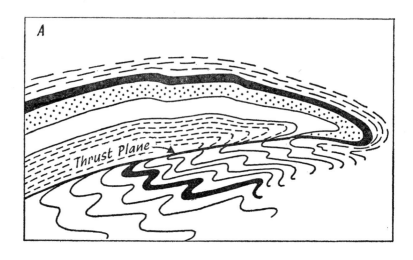

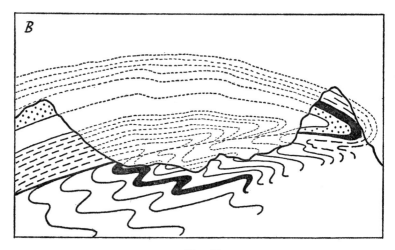

FIG. 56

Nappe structures in the Alps. *A*: a great pile of sediments, folded and pushed over from the left, has slid along a thrust plane over contorted rocks. *B*: the same 'nappe' is seen after a long period of erosion, when part of the front of the structure has been separated as a 'klippe' from its roots.

which form the great wall of mountains on the north side of the Inn Valley in the Austrian Tyrol, are thought to have travelled in this way. If so, they must have been pushed right over the older nappes which form the high mountains in the south-west corner of Austria, from a 'root zone' in Italy.

The idea of such far-travelled rocks was not at first accepted by non-Alpine geologists, and even among that remarkable fraternity there is considerable disagreement about how far a particular mountain has travelled.

The simple explanation of the great compression which produced the Alps is that the African plate moving north collided with and over-rode the European plate. In this case, therefore, we have two continents colliding (as shown in Figure 57) rather than the somewhat

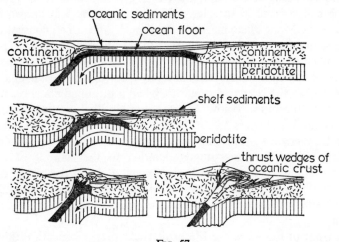

FIG. 57

Collision of two continental plates, involving both subduction and obduction of ocean floor material (after Dewey and Bird).

simpler situation, described and illustrated in Chapter I, of a collision between a continent and an ocean. Thus all the great folds of the Alps were pushed over to the north until the last stage in the movements, when the Dinaric Alps buckled back to face the other way (see Figure 58).

The same thing is seen much more clearly in the Himalayas, where there is one great set of folds and thrusts pushed towards the north and another towards the south, with the high plateau of central

Tibet in between. In this case it was a collision between the Asian plate and India heading north from Gondwanaland.

Returning to the Alps, we can see that as fast as the nappes were piled up, so they were being rapidly worn down again. Tremendous thicknesses of coarse sediment accumulated in front of the rising mountains. These are known generally as the **Molasse** and yield fossils ranging in age from Upper Oligocene through the Miocene. They consist mainly of soft sandstones and conglomerates and are worked locally for a dull grey concrete-like building stone such as is used in Mozart's Salzburg. In places, the later Alps have come to rest on the Molasse and so demonstrate that the orogeny was a very lengthy business.

Here and there, sticking up through the folded Mesozoic and Caenozoic rocks of the Alps, there are rigid masses of older rocks which had suffered in the Hercynian Orogeny. The best known of these is the Mont Blanc massif in the French Alps which consists of a large granite intrusion and associated metamorphic rocks. This and other ancient 'massifs' stood up like breakwaters amidst the waves of Alpine folding. They fractured but did not bend in the same way as the later rocks, and they had a profound effect on the resultant structures. Below Mont Blanc, in the mountaineering capital of Chamonix, is one of the few memorials to a geologist that have been erected by an ungrateful world. The first of the Alpine geologists—de Saussure—stands pointing to the summit of his highest mountain. It was he who first pointed out, at the end of the eighteenth century, the importance of studying modern mountain ranges such as the Alps in order to understand the history of the earth.

In front of the Alps, the Mesozoic rocks plunge beneath a thick cover of Molasse which forms the central plain of Switzerland and beyond this they rise again in the Jura Mountains. One can see this clearly in a boat trip down the length of Lake Geneva. Starting from the eastern end one has picture postcard views of Byron's Château de Chillon against a background of the foremost Alpine ranges; then before Lausanne, one enters the park-like Molasse plain and finally as Geneva and its high fountain come into sight at the west end of the lake, one sees the first regular ridges of the Juras.

The rolling, forested, wine-growing countryside of the Juras gives its name to the middle period of the Mesozoic era and is chiefly formed of gently folded Mesozoic limestones. The Juras swing round

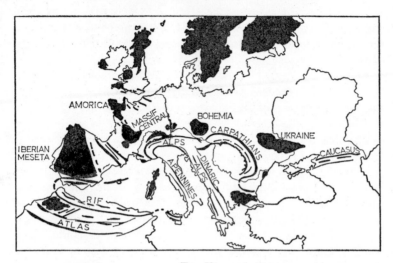

FIG. 58

Alpine folding in Europe: map showing the main belts and axes of folding during mid-Tertiary times. The old stable masses are shown in black, the fold-belts are shown as heavy black lines.

in a broad arc from the old dukedom of Burgundy in south-east France, through Switzerland into Germany. In them the force of the Alpine folding is considerably diminished. The rocks are disposed in a series of parallel anticlines and synclines which seem to have slipped on a surface of Triassic salt deposits, like a table-cloth on a highly-polished table. The folding was restrained at each end by ancient rigid blocks (see Figure 58). At the south end there is the Massif Central west of Lyon and in the north there are the old rocks of the Vosges and the Black Forest in Germany. Between these two breakwaters, the folds of the Juras have spread out like waves in front of the storm centre of the Alps. Beyond this again minor ripples spread on northwards as far as southern England.

The most intensive Tertiary folding in England is to be seen in the Isle of Wight and on the adjacent mainland of Hampshire and Dorset. In the hills at the south end of the delightful Victorian island, the Chalk dips gently southwards towards the sea. Northwards, however, it turns down almost vertically to form the island's backbone of hills. In front of the Chalk, Eocene and Oligocene rocks are also steeply inclined, so there is no doubt as to the age of the move-

199

ments. At the western end of the island, the Chalk runs out to sea in the line of sea-stacks known as the Needles, and this is continued on the far side of the Solent by the Old Harry Rocks off Ballard Down.

At Lulworth Cove in Dorset, Upper Jurassic rocks can also be seen involved in the folding, and in the marine excavation known as Stair Hole, just west of the main cove, quite an Alpine-style crumple can be seen in the Purbeck limestones.

Elsewhere in southern England, the Alpine folding has shaped the main features of the countryside. The Weald of south-east England is a great anticlinal structure produced at this time, and the London Basin is the complementary syncline to the north. The anticlinal folds are all along east-west lines and tend to be steeper on their northern sides. This is particularly well shown by the Chalk in the Isle of Wight, but can also be seen in the Weald. Thus the 'Hog's Back' at Guildford in Surrey is formed by steeply-dipping Chalk on the north side of the Wealden structure. In the area of the Weald and the Hampshire Basin, the folding is all within the Mesozoic rocks, for the Palaeozoic floor is very deeply buried. Under London, however, the minor structures in the later rocks appear to result from faulting in the Palaeozoic foundations just beneath.

The main lines of Alpine folding in England are shown in Figure 58. Their east/west orientation may cause them to be confused with the Hercynian folds farther west, but of course they affect much younger rocks. The older rocks of 'Highland Britain'—in the north and west—were not much affected by the Caenozoic earth-movements, though there were some 'posthumous' movements along old fault lines. At about the same time, Britain must have acquired the generally south-easterly tilt which gives it the scenic pattern it has today. As a result of this tilt, the main outcrops in England run from north-east to south-west and a traverse from Anglesey to London takes one up the stratigraphical column from the Precambrian to the Caenozoic over a regular series of escarpments and dip-slopes.

Before leaving the Alpine Orogeny, a word more should be said about a matter discussed in earlier chapters—the possible connexion between orogenies and organic evolution. The Alpine Orogeny in Europe is the one closest to us in time and probably the one of which we have the fullest knowledge, but if we compare the faunas of the Palaeogene with those of the Neogene, we can see nothing that could be called a major change. The most impressive and rapid evolutionary changes during the Tertiary were those that went on in the mammals, but we cannot really say that the evolution of the horse or the elephant was speeded up or affected in any way by the mountain-building movements. Perhaps the only startling event in the organic world which occurred during this general period, was the emergence of man himself, but this seems to have happened far away from the Alpine mountains, and in any case was just one more species.

It may be that we are too close to the Alpine Orogeny in time to be able to observe its effects. In the same way an historian cannot write a fair and balanced account of an event and its after-effects in less than fifty or a hundred years after it happened. We must wait for a few million more years before we can be sure what really happened in the Alpine Orogeny.

NEOGENE SEDIMENTS

The Alpine Orogeny fundamentally altered the structure of Europe, and the sedimentation which accompanied and followed the orogeny was very different from what had gone before. Generally speaking there was uplift towards the end of the Oligocene Epoch, and Miocene sediments are found in a series of new basins in front of the mountains.

The Molasse of the Alps has already been mentioned. This began as a land or freshwater deposit in a series of separate basins, but later the sea broke in from the east and west and finally it retreated again. Probably the most famous Miocene freshwater deposit is that at Oeningen on the shore of Lake Constance. This has yielded a wonderful collection of plants, insects and vertebrates. The plants indicate a warm, humid climate comparable to that of the Canary Islands at the present day, and seem to have been deposited in regular layers recording the different seasons of the year. One of the first vertebrates found here (long before the days of geology) was thought to be the skeleton of a sinner drowned in Noah's flood. It was therefore called '*Homo diluvii testis*', but is now known to be a giant salamander.

Thick marine deposits of Miocene Age were especially well developed in the Vienna Basin in eastern Austria. The sediments are highly fossiliferous and, like similar Tertiary deposits in many other parts of the world, they contain vast quantities of oil. The recovery of this oil is now one of Austria's most profitable industries. Oil occurs in strata of different ages in different places. In Britain, the only oil accumulations of importance that have so far been discovered on-shore are in Carboniferous strata. In Canada it is recovered from Devonian reefs, in Texas from thick Permian sediments and so on up the stratigraphical column. But the Tertiary rocks seem to be particularly favoured, and some of the most profitable oilfields in the world take their oil from these later rocks. The origin of crude oil in the rocks is still a matter of dispute, but it is almost certainly of organic origin and always seems to be associated with an abundance of past animal life. Since it is liquid, it flows through porous rocks and may be found far from the sediment in which it originated. It also tends to seep out at the surface and be lost, so it is necessary in oil exploration to find geological structures such as anticlines where the oil may have been trapped beneath an impermeable layer. Many deep borings are put down in oil exploration and a great deal depends on the identification of specimens from boreholes by means of the fossils which they contain.

Towards the end of the Miocene, the various basins in front of the Alps became cut off from each other as separate inland seas. Besides the Vienna Basin, there were also basins in Rumania (another oil basin), Hungary and southern Russia. As rivers flowed into these enclosed seas the water became brackish and eventually fresh again. These changes in salinity had some interesting effects on the animals living in the basins, which either became extinct or evolved very rapidly to adapt themselves to the changed conditions. Some of the unusual brackish water forms still survive in the Caspian Sea, which is the last remnant of the south Russian basin.

In the Pliocene epoch which followed, the sea retreated for the last time from the central part of Europe and marine deposits are restricted chiefly to the southern margins of the continent, in Italy, Greece and Spain. They are particularly well developed in Italy where they cover the greater part of the country and yield a fauna not very different from that of the present Mediterranean. In the same region at this time there began another volcanic episode which is still going on. Perhaps connected with the formation of the Mediterranean as it is today, was the development of weaknesses in the crust and the eruption of the volcanoes of Italy, Sicily and other

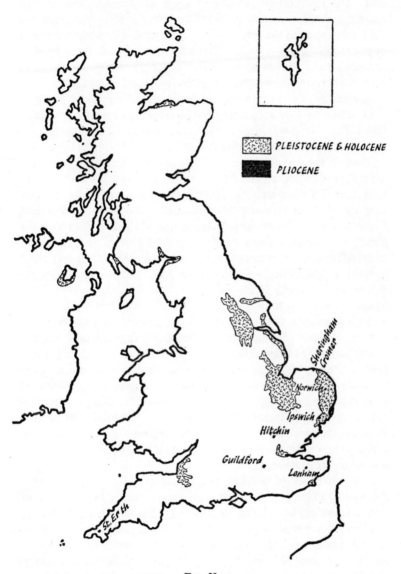

FIG. 59

Map showing the main outcrops of Neogene and Quaternary rocks in Britain, excluding glacial deposits.

nearby islands. These effects reached their maximum in the Pleisto-cene. The intermittent activity of some of the volcanoes is well known, and there are also quite sudden up and down movements of the land relative to the sea. Thus Pliocene marine deposits are found in places in Italy more than 3,000 feet above the present sea-level.

Elsewhere in Europe Pliocene deposits were chiefly laid down in lakes and rivers, though there was an important extension of what is now the North Sea into eastern England and the Low Countries.

In Britain we have only a very poor sedimentary record of Neogene times. There are no Miocene deposits at all and the Pliocene is only represented in thin and scattered patches (see Figure 59). The chief of these is the **Coralline Crag** near Ipswich in Suffolk. During the nineteen-forties the author helped to dig slit trenches in Sudborne Park near Orford, which were fortunately never needed for military purposes, but for several years afterwards yielded vast numbers of fossils to scores of collectors. 'Crag' is a local East Anglian term for poorly-consolidated shelly sands. The first part of the formation name refers to the presence of abundant very small colonial organisms formerly called 'corallines'; they are in fact **bryozoans** and quite un-related to the corals. There are several later formations, also called 'crags', which used to be included in the Pliocene (and will be found as such in many textbooks). Nowadays, however, they are regarded as of early Pleistocene Age and will be dealt with in the next chapter.

The fauna of the Coralline Crag is truly remarkable (see Figure 55); there are many different types of fossils, but bivalve molluscs (or lamellibranchs) predominate. The fauna included many forms of southern affinities. In the later 'crags' these disappear and are re-placed by northern forms, indicating a sharp deterioration in climate. In places the sands have been cemented sufficiently by lime and iron compounds to form a rock-bed which can be quarried and used as a local building stone.

A whole series of smaller, but in some ways more interesting deposits of Pliocene Age occur high on the North Downs in Kent between Ashford and Folkestone. They are found as irregular masses of loose reddish sand filling solution hollows in the top of the Chalk. Near Lenham they have yielded quite a large marine fauna which indicates an early Pliocene Age and can be correlated with faunas in similar sands on the other side of the Channel (for example near Antwerp in Belgium). The interest in these deposits lies in the fact that they imply a submergence of eastern Kent in geologically recent times and its elevation again so that these Pliocene marine deposits are now found up to 600 feet above sea-level.

But there is more to it than this, for it now seems that the whole of south-east England may have behaved in the same way. In the absence of fossils, another method has been used to correlate Pliocene sediments from place to place. This is the careful study of the mineral grains which occur in the sands. It has been shown that sands with the same unusual minerals occur all along the North Downs and even on the Chiltern Hills across the other side of the London Basin. These sands and associated gravels rest on a fairly well-marked platform high up on the Chalk hills and certain geologists claim that this platform was cut by the waves of a Pliocene sea which covered much of south-east England. If this is true, it is a remarkable demonstration of how small a record may be left by a major geological event. Here we have a period of subsidence followed by a widespread marine invasion with erosion and deposition and then by re-elevation to a considerable height, and the only record of all this is a few patches of sand which will probably be worn away in a few hundred years. It makes a good geologist all the more sceptical of the generalizations which are made about earlier periods.

Further slightly conflicting evidence comes from two places where fossils have been found in disturbed blocks of similar sediments, one at Netley Heath on the North Downs near Guildford and the other at Rothamsted on the Chilterns north of St. Albans. The fossils are similar to those of the 'Red Crag' (see Chapter XI) which used to be called Pliocene but is now regarded as early Pleistocene. This does not fit in with the evidence of the Lenham fossils and many geologists would not accept a marine invasion of this extent in early Pleistocene times.

A well-marked platform, about 430 feet above sea-level, can be seen over much of Cornwall and in Devon. It is very clear, for example, on the Lizard peninsula, where it bears supposedly marine gravels. A similar platform truncates the ancient rocks of South Wales, notably in Pembrokeshire and on the Gower Peninsula (see Plate 17). All these places are characterized by an almost monotonously flat inland landscape which ends abruptly with sheer cliffs dropping straight to the sea. This platform was obviously cut by the sea and near Camborne in Cornwall, the small Carn Brea granite boss clearly stood up as an island, with its base still surrounded by old beach shingle and marine clays and sands. The best evidence for

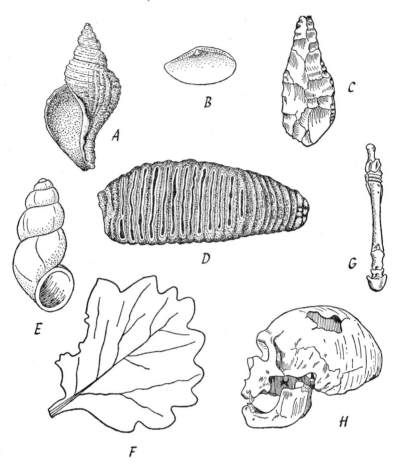

FIG. 60

Quaternary fossils. *A*: 'left-handed' marine gastropod (after Institute of Geo-
logical Sciences), $\times \frac{2}{3}$; *B*: cold water lamellibranch, $\times \frac{1}{2}$; *C*: early man-made
hand-axe (after Woodward), $\times \frac{1}{3}$; *D*: grinding surface of a mammoth tooth,
$\times \frac{1}{6}$; *E*: minute land snail (after Kerney), $\times 15$; *F*: leaf of the dwarf birch (after
Walton), $\times 8$; *G*: fore-limb of the true horse, $\times \frac{1}{16}$; *H*: skull of Neanderthal
man, $\times \frac{1}{5}$.

the age of this surface comes from St. Erth in Cornwall, where
deeper water muds with abundant Pliocene fossils accumulated in a
hollow on the sea-floor.

XI

THE ICE AGE

This chapter deals with the remarkable events of the Pleistocene Epoch, when great sheets of ice spread over much of the northern hemisphere.

I f one goes to any of the touristic glaciers of the Alps it is obvious, even to the least observant, that the ice must have formerly ex-tended much farther than it does today. All down the valley below the glacier's mouldering snout there are the signs of ice action; the valleys sides are scraped and scratched down to the bare rock and there are great untidy heaps of rubbish left by melting ice. Such observations were made many years ago by Swiss geologists and they came to the general conclusion that in the geologically recent past, the glaciers of the Alps were much thicker, more extensive and more continuous than they are today. There was in fact a single great capping of ice all over the central Alps.

In 1840, Louis Agassiz, the great Swiss palaeontologist who was the chief protagonist of these ideas, came on a visit to Edinburgh. He was taken to see the local geology (mainly Carboniferous igneous rocks). On the outskirts of the city, he stopped and pointed to a smoothed and scratched rock surface which can still be seen today. Remembering the walls of his glacier valleys in Switzerland, he said, 'That is the work of land ice,' and so started one of the fiercest con-troversies of British geology in the last century. From the recognition of massive ice action in Scotland came the notion that the greater part of Britain had at one time during the recent geological past, been covered by a sheet of ice hundreds of feet thick. Such ideas at first received considerable opposition, even from such notables as Sir Roderick Murchison, but the picture which finally emerged was even more surprising. It now appears that an ice sheet, centred on Scandinavia, at one time spread down far over the north German plain and swept across the North Sea to meet British ice-sheets which reached as far south as the line of the Thames and the Severn (see Figure 61).

There were separate smaller ice-caps on mountainous areas such as the Alps, the Pyrenees and the Caucasus. In North America at the same time, a huge ice-sheet covered the whole of Canada and a long lobe of ice extended way south of the Great Lakes. What is more, this did not just happen once, but in four distinct **glaciations**, separated by **interglacials** when the climate grew warmer. The whole Pleistocene Epoch, with its four glaciations, did not last more than 2 million years, which is a very short period of time to a geologist. It is probably only about 10,000 years since the last ice-sheet began to retreat and we cannot yet be sure that we are not living in another interglacial.

The word 'glaciation' and the mention of modern glaciers is in a way misleading, for the ice-sheets of the Pleistocene were on a vastly greater scale than anything we can see in our familiar mountain

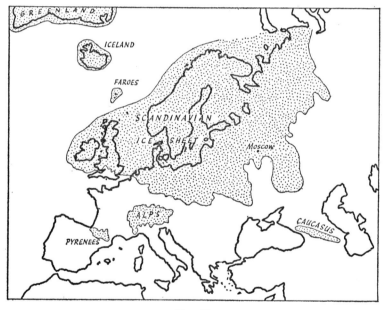

FIG. 61

Map showing the maximum extension of the ice-sheets over Europe during Pleistocene times. (Simplified from *A Palaeogeographical Atlas of the British Isles* ... by L. J. Wills, by kind permission of the author and Messrs. Blackie & Son, Ltd.)

ranges today. They are more comparable to the cappings of ice which survive in Antarctica and Greenland and which together constitute something like 96 per cent of the area at present covered by ice.

The greatest problem of this 'ice-age' (and the earlier ones) is why they happened. The geologist is usually content to demonstrate that they did happen and then to leave it to the physicist to suggest a mechanism. This is convenient for geologists because no completely satisfactory mechanism has yet been found. Some look for the explanation within the earth and its atmosphere, others look outside it. Certainly glaciations are unusual events in earth history and they seem to coincide with periods of continental emergence—when there was a larger area of land and a smaller area of sea than usual.

To form a glacier it is necessary for snow to accumulate from year to year and not to melt during the summer. The level at which this happens rises from sea-level in the polar regions to nearly 18,000 feet in the tropics. Mount Kenya in East Africa has about ten glaciers although it is very nearly on the equator. Obviously then, a rise of the land surfaces is likely to produce more extensive glaciers, but it cannot be the whole explanation.

Some physicists have suggested changes in the amount of heat reaching the earth from the sun, either through changes in the sun itself or in the relationship of the two bodies. Others have suggested a reduction in the amount of solar heat reaching the earth due to the intervention of dust clouds, either in the atmosphere or in outer space. There are almost as many theories as theorists.

Glaciers and ice-sheets act as geological agents in two ways. In upland areas they erode the landscape in their own peculiar way, and in lowland areas they deposit sediment of distinctive types in distinctive patterns.

Ice erosion differs markedly from water erosion and produces a different type of scenery. The jagged mountain ranges of the world, with their peaks, knife-edge ridges and sheer cliffs are the result of ice action. Water enters cracks in the rocks, freezes and expands, and so rock fragments are constantly being split off. The sharp ridges and peaks which delight mountaineers are formed by the glacial valleys biting back into the mountains and meeting one another. The head of a valley glacier is a small basin or **cirque**, where the snow originally accumulated to form the ice of the glacier. The highland areas of Britain abound in cirques which have long lost their glaciers. A famous example in North Wales is Llyn-y-Gader below the summit of Cader Idris (see Plate 3). Such cirques nowadays usually contain a small lake.

As the ice moves downhill under gravity, it smooths and scratches the surfaces over which it passes. It is helped in this by the rock fragments carried along in the ice. When a glacier moves down an old river valley, it considerably modifies the valley's shape. The broad, slow-moving ice tears away at the valley floor and converts what was previously V-shaped in cross-section to a flat-bottomed U-shape. At the same time, a glacier does not meander like a stream, but straightens its valley as much as possible by cutting away projecting spurs. Also, the main glaciers erode their floors much more rapidly than tributary ones, and when water once more replaces the ice, the side streams are left in **hanging valleys** high above the main stream, and often plummet down as spectacular waterfalls.

Dobb's Linn in the Southern Uplands, where Lapworth did his famous work on the graptolites (see Chapter III) is an excellent example of a glacial valley, but there are hundreds of others all over Scotland. The long highland lochs are found in valleys which were greatly deepened by ice action. The fjords of Norway are steep-sided valleys which have been gouged out by ice and later flooded. Mention has already been made of the smoothing and scratching action of moving ice. Hard rock surfaces are often quite highly polished in this way and the scratches are often a useful indication of the direction of ice movement.

The smoothing and 'plucking' effect of the ice often produced groups of small asymmetrical knobs of bare rock which are called **roches moutonnées**, perhaps because of their resemblance to flocks of resting sheep (see Plate 18). The same effect on a grander scale is seen in Edinburgh, where the Castle Rock stood up and was plucked and steepened by the passing ice, but protected the softer sediments which form the gentle slope through the old city down to Holyrood Palace.

The deposits left by moving ice are many and various. All such deposits are known under the general name of **drift**. The rock fragments carried along by ice are not rolled and rounded as they would be in water, but they may be smoothed, scratched or ground to fine powder or 'rock flour'. The extreme limit of a glacier or ice-sheet is the point at which it melts away as fast as it advances. At this point, the debris carried by the ice accumulates in an irregular ridge or

moraine. If it then retreats in a spasmodic manner, it may leave a series of such ridges. In the mid-western states of the U.S.A. there are hundreds of miles of dead flat plains interrupted only by low morainic ridges. These ridges are often scores of miles in length and form gentle arcs, lying one within another, and marking the successive halt stages in the retreat of the ice.

The material which makes up moraines is **'boulder-clay'** or **till** (described in Chapter VII). This is an unstratified deposit consisting of boulders of many sizes and many types, embedded in a fine-grained clayey matrix. Moraines are only minor accumulations of boulder-clay at the extreme limits of the ice advance or at successive retreat positions. Much greater quantities are found spread over the vast areas where the ice-sheets melted in position. This is sometimes called the 'ground moraine' as distinct from the 'terminal moraine' and may be hundreds of feet thick. In places boulder-clay is heaped up in streamlined hillocks or **drumlins**, which have a blunt, steeper side towards the source of the ice and a gentler slope the other way.

Apart from the deposits laid down by the ice itself, there are also the sediments carried away and laid down later by melt waters. These have the characters of ordinary stream deposits of sand and gravel, but their material is derived from the ice and not from local rocks.

One of the most fascinating studies for glacial geologists is tracking down the source of boulders found in the drift. Ice can carry large blocks for considerable distances, and when it melts it often leaves them in unlikely places. There are many famous 'rocking stones' in Britain and elsewhere, which are **erratics** or far-travelled boulders which have been left delicately balanced by the melting ice. Erratics are so called because they are completely out of place where they are found. Thus great erratics of metamorphic rock from the Alps are found perched on the limestone hills of the Juras, and in the American Mid-West there are chunks of gneiss scattered everywhere which must have come from the Canadian Shield. If the source of the erratics can be recognized exactly, then this is a clear indication of the direction of ice movement.

The Shap Granite of Westmorland is an easily recognized rock, full of large pink felspar crystals. Erratics of Shap Granite are found in a long trail extending right across the Pennines as far as the Yorkshire coast. This records the passage of a tongue of ice moving east from the Lake District. Even more spectacular erratics are those

211

from Norway found in the boulder-clays of East Anglia and even picked up as far south as Finchley.

The Mecca of all British Pleistocene geologists is East Anglia, where there is a fine record both of normal sedimentation and of the deposits and action of ice. Nevertheless, the story here is far from clear and the experts are not agreed on some of the main chapters, so that it is difficult to write about geologically recent events in this part of England without denying someone's favourite theory.

One of the most difficult problems about Quaternary geology has been deciding when it started. For years there has been uncertainty about what should be called Pliocene and what Pleistocene. This is not just an argument about words, but a real problem, because the usual criteria of correlation do not seem to apply. This is partly due to the confusing effects of a deteriorating climate and partly because of a lack of suitable rapidly-evolving fossils. In spite of these difficulties, some agreement was reached at the International Geological Congress in London in 1948. A standard section was then chosen in Italy where the base of the Pleistocene can be clearly defined in marine sediments away from the immediate effects of the glaciations. The correlation of this section with East Anglia resulted in most of the 'crags', previously called Pliocene, becoming Lower Pleistocene. This fits in with climatic criteria, for the first marked influx of cold-water forms came immediately after the deposition of the Coralline Crag.

The first sediments of the Pleistocene in Britain are therefore the **Red Crag**. This succeeds the Coralline Crag laterally rather than vertically and there is evidence of considerable erosion between the two. The Red Crag extends south to Walton-on-the-Naze and probably nearly to Framlingham in the north. As it name implies, it is predominantly red in colour, due to staining by iron oxides, and it is nearly always highly fossiliferous (see Figure 60). Its broken shallow-water shells and false-bedded sands show that it was laid down close to a shoreline.

At the bottom of the Pleistocene Crags in places is an unusual deposit, a foot or two thick, of bones, teeth, far-travelled boulders, derived fossils and phosphatic concretions. The last-named were dug in the past for use as a fertilizer. There are large sharks' teeth, whale bones and some lumps of sandstone with Oligocene or Miocene fossils which may have come from an unknown outcrop in the North Sea. Some geologists have suggested that the non-local rocks were brought by floating ice, and others have correlated the appearance of northern fossils in the Red Crag with the first glaciation on the continent. This is far from being generally accepted, but there is said to be some evidence of a 'warming-up' of the fauna in the next deposit—the **Norwich Crag**.

At Aldeburgh brickyard on the Suffolk coast, the Norwich Crag is seen resting on the Coralline Crag of the Pliocene. It is a close inshore deposit of shelly sands and contains drifts of freshwater shells and land vertebrates in places. This deposit is thought to extend over much of Norfolk, but is hidden by glacial deposits. Its name comes from the excellent exposures which were formerly available around the cathedral city. At the base of the Norwich Crag hereabouts, there occur chipped pieces of flint which have been regarded by some archaeologists as primitive human tools. Even earlier and cruder supposed tools have been found in the bone bed at the base of the Red Crag. It is almost impossible to prove whether a particularly roughly-chipped stone was produced by human intent or natural accident, and the amount of literature and heat generated by these matters is immense.

The next formation in the East Anglian Pleistocene succession is a series of clays and silts called the **Chillesford Beds**. They are found at a number of scattered localities apparently resting on the Red or Norwich Crag. If these localities are linked up on a map, they are found to lie along a sinuous line from the type locality of Chillesford near Orford, northwards to the Norfolk coast. This sinuous line was interpreted as a distributary of the then combined drainage systems of the Rhine and the Thames. The beds were thought to have been laid down in a quiet deltaic estuary, and a famous find at Chillesford was the complete skeleton of a whale which had drifted into the estuary to die. If this story is true, it provides a wonderful example of a 'fossil river' and has been seized upon as such by many textbook writers. As is usual in geology, however, the beautiful theory was followed by scepticism, and some recent workers think that the Chillesford Beds are just a local variation of the Norwich Crag, at different levels and of no particular significance.

The next formation is the **Weybourne Crag**, seen on the north Norfolk coast and containing the first British representatives of certain cold-water shells. Some correlate this with an advance of the ice fronts. It is followed by the most interesting of all these deposits—the **Cromer Forest Bed**. This includes both marine and freshwater fossils, but is chiefly famous for its drifted trees and other plant remains, and for its large vertebrate fauna. Many (probably too many) species of deer have been described from this bed, and other fossils include the mammoth and the 'sabre-toothed tiger'. Conditions were probably like those of Norfolk as it is today, but a mixture of fossils resulted from their remains being washed down rivers into a deltaic region.

The earliest undisputed human artifacts of East Anglia are found in the Cromer Forest Bed. They are large hand-axes and other tools which belong to the earliest of the culture divisions recognized by archaeologists (see table below).

HUMAN ARTIFACT CULTURES IN THE QUATERNARY PERIOD

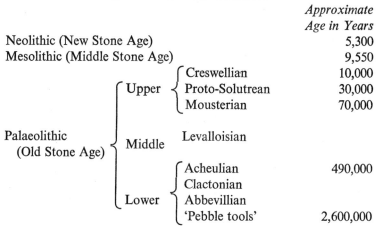

				Approximate Age in Years
Neolithic (New Stone Age)				5,300
Mesolithic (Middle Stone Age)				9,550
Palaeolithic (Old Stone Age)	Upper	Creswellian		10,000
		Proto-Solutrean		30,000
		Mousterian		70,000
	Middle	Levalloisian		
	Lower	Acheulian		490,000
		Clactonian		
		Abbevillian		
		'Pebble tools'		2,600,000

Almost immediately on top of the Cromer Forest Bed in the cliff sections of north Norfolk, comes the first deposit which can be directly attributed to the action of ice. This is the first of several boulder-clays and is largely made up of local Mesozoic and Tertiary material, but also includes many derived blocks from Scotland and Scandinavia. Two of the most distinctive of these are a **rhomb-porphyry** (which is a reddish volcanic rock with large diamond-shaped crystals of felspar) and **laurvigite** (a dark shimmering plutonic rock of attractive appearance, much used on 'Woolworth's' and other shop fronts). Both of these rocks are well known in their un-disturbed occurrences in southern Norway and could not possibly be confused with any British igneous rocks, so they provide clear proof that the ice-sheet which laid down this boulder-clay travelled right across the North Sea.

This particular ice-sheet probably did not travel far inland, for farther south, instead of boulder-clay, there are sands and gravels deposited by melt-water streams. Above the boulder-clay in places there are sands with temperate shells which are interpreted as inter-glacial deposits. There is also evidence of considerable erosion before the next glacial deposit. This used to be called the 'Chalky Boulder-clay' because of the abundance in it of Chalk fragments from inland England.

The earlier ice-sheet had laid down its load peacefully and without disturbance of the underlying sediments. This was not so with the ice-sheet which produced the second boulder-clay. It ploughed into the sediments, pushing them aside and deforming them in a manner only comparable with the tectonic distortion of old rocks in the Highlands. Movements within the ice-sheet itself produced fantastic contortions in the boulder-clay which was left when the ice melted. In many places it has incorporated the earlier boulder-clay and has considerably affected the solid rocks beneath. At Trimingham on the Norfolk coast, a dome-like structure was produced in the top of the Chalk and this reveals the youngest Cretaceous zone known in Britain. Not far away near Overstrand there is at present exposed (though much neglected) a great overturned fold in the Chalk carrying with it the Weybourne Crag. Even structures such as inverted false-bedding can be seen in these Pleistocene sediments as a result of ice pressure. Perhaps the most remarkable feature of all is along the coast east of West Runton, where a thick bed of Chalk, about a a quarter of a mile long, has been levered up and enclosed within the boulder-clay.

Early in the present century, a classic study in the Alps first re-vealed the existence of four distinct glaciations during the Pleisto-cene. This discovery considerably affected later thought on glacial matters, and geologists went to work expecting to find evidence of four glaciations. In the words of a famous French geologist (re-cently quoted by Sir Edward Bailey in another connexion), 'To see things it is necessary to believe them possible.' The four glaciations were duly recognized in East Anglia, but have since been much dis-puted by later workers. It would certainly be difficult to fit in four glaciations if, as now seems possible, a glaciation was going on elsewhere while the Crags were being deposited in Norfolk and Suffolk.

One explanation may be that some of the episodes which were formerly taken for distinct glaciations were in fact only phases in a single glaciation. We know now that every advance of the ice was accompanied by slight backward and forward oscillations, and during a single glaciation, ice-sheets may advance on an area from different directions.

Returning once more to the lobe of ice which spread south from the Great Lakes, here no less than *seven* oscillations have been recognized in the last of *four* phases of the last glaciation. About half-way down the west side of Lake Michigan is a fossilized spruce-bog (sub-Arctic by modern standards) where the trees were actually pushed over and snapped off by the farthest advance of this ice. Growth-rings in the trees show that they had a difficult time during the last ten years or so of their existence. Trees are snapped off in the same way by minor advances of glaciers in Alaska today. What is particularly interesting about this locality is that the event can be dated almost exactly. In the last few decades a method has been developed for dating such recent geological events by using radio-active carbon. This is essentially the same method as was described in Chapter II for uranium, but radio-carbon has a very much shorter life. Its great advantage is that it enters into all living matter and so gives a date for an organism's death. This method has been used to date papyrus from Egyptian tombs and the famous Dead Sea Scrolls. The results fit in very well with the archaeological evidence. In the case of the Lake Michigan locality, the trees appear to have died about 10,600 years ago. The ice which killed them only went a few miles farther and then began its retreat.

So far, we have only considered areas which were actually glaciated, but there were also wide regions which, though not covered by the ice, were greatly affected by its presence. Immediately in front of the ice-sheets there was a belt where the ground was usually frozen and nothing could grow except mosses, lichens and an occasional dwarf birch (see Figure 60). This is comparable to the modern tundra region, and it received much of the outwash material from the melting ice. Beyond the tundra there is today a broad steppe region with hot summers and cold winters. During Pleistocene times a steppe-like region extended right across central Europe and Asia. It is characterized particularly by a deposit known as **loess** which is fine wind-blown dust. Loess crumbles easily in the hand and being wind-blown is completely without pebbles. It is particularly well developed in central China where it forms a very fertile soil. An early European visitor there wrote home complimenting the Chinese on their industry in having removed all the stones from their soil! In southern Britain there are deposits of what is known as 'brick-earth' (from its use for brick making); this approaches loess in composition. A deposit of this nature at Pegwell Bay in Kent has been shown to be a true loess, but generally speaking this type of sediment seems to have been developed mainly on the broad low-lying plains around major rivers such as the Mississippi and the Yangtze Kiang.

Beyond the steppe there comes the belt of deciduous forests which would cover England today were it not for the activity of men. All of these belts with their distinctive faunas and floras, moved backwards and forwards across Europe in front of the advancing and retreating ice fronts. During the glaciations they were far to the south, and in Europe many animals and plants met the impassable barrier of the Alps and became extinct. In North America on the other hand, where the mountains run north and south instead of east and west, the warmth-loving organisms were able to move on south and come back as the ice retreated. So it is that the flora of North America is today more varied than that of Europe.

There is some evidence to indicate that in one interglacial at least, conditions in England were warmer than they are at present. A clear picture of life and conditions during the last interglacial period has recently been obtained from excavations in Trafalgar Square, right in the centre of London. Elephants' teeth and other large bones have been found hereabouts for many years, but only in 1957, when the foundations were being dug for the new Uganda House, was the fossiliferous deposit properly studied. Bones, shells, insects and plants were all found in large numbers in the silts and sands of an earlier River Thames. From these emerged a picture of a swiftly-flowing river, bordered by marshes and open parkland scenery. The climate was warmer than it is at present and the larger animals of the neighbourhood included the straight-tusked elephant (contrasting with the woolly mammoth of the glacial epochs), lions, bears, hippopotami, rhinoceroses and various species of deer.

One animal which flourished particularly on the open grasslands of the late Pleistocene was the so-called 'Great Irish Elk' with a tremendous spread of antlers. This is particularly common in the post-glacial peat-bogs of Ireland, but although it survived the glaciation it became extinct before historic time. The interesting suggestion here is that it became extinct when the deciduous forests spread northwards for the last time, and it could not get its huge antlers between the trees!

One effect of the Pleistocene glaciations was the raising and lowering of the relative sea-level. Two processes were involved, one resulting in the rise or fall of the sea and the other in the rise or fall of the land. During the glaciations, great quantities of sea-water were tied up in the ice-sheets and so the sea level fell, to rise again in the interglacials when the ice melted. On the other hand, in the areas which were actually glaciated, the sheer weight of the ice caused a depression of the land from which it slowly recovered during the interglacials. The former process was probably world-wide in its effects and these are most clearly seen around the Mediterranean, where there is hardly any tidal rise and fall to confuse the issue. The effect of the weight of the ice is best seen in Scandinavia, the centre part of which has been rising steadily ever since the weight of ice was removed. Even parts of Scotland have risen more than 100 feet since the glaciations.

Changes in sea-level are shown by various features. High sea-levels are revealed by **raised beaches**. These are the platforms cut by the sea often way above its present high-tide level and still covered with the sands and shingles of the original beach. The latter sometimes contain shells and other fossils which suggest temperate conditions and therefore an interglacial age. Around the English coasts there are well-marked raised beaches at about 100 feet above Ordnance Datum (for example between Chichester and Arundel), at 60 feet (for example at the south end of Portland Bill) and at 25 feet (for example on Hope's Nose near Torquay).

It can be shown that the lower raised beaches are progressively later in age and we can trace the whole process back to the high level surface cut by the Pliocene sea, which was discussed in the last chapter.

Low sea-levels in the Pleistocene are more difficult to recognize, but at several places around the English coast **submerged forests** may be seen at very low tide. These are the remains of trees, still in their position of growth, which were drowned when the sea rose again. Trawlers often bring up peaty deposits and plant remains from the Dogger Bank in the North Sea, sometimes from nearly 200 feet down. This is evidence of a very much lower sea-level (relatively speaking) in Pleistocene times, when the British Isles were joined to the continent and the Thames and the Rhine flowed together to an estuary far to the north.

Such studies of old sea-levels are difficult and complicated because of the conflicting factors involved. There is also the danger that they may have been modified by later tectonic earth-movements. The classic case of such movements in recent times is provided by the so-called 'Temple of Serapis' at Pozzuoli near Naples. Here three pillars of what was in fact a Roman market-place clearly show the borings made by marine organisms when they were deeply submerged in sea-water. Sir Charles Lyell chose these pillars as the frontispiece of his classic work 'Principles of Geology' and they appear on the Lyell medal of the Geological Society of London. They were his outstanding demonstration of the rise and fall of sea-levels within historic times and these movements can be roughly correlated with the intensity of volcanic activity in the neighbourhood.

In spite of all that has been said so far, it is important to remember that many parts of the world at this time did not see so much as a snowflake. All the normal geological processes and phenomena were taking place as before. In the western part of the U.S.A., there was quite an important orogeny going on and great thicknesses of detrital material were accumulating on the 'Great Plains'. The Apennine mountains in northern Italy were still rising and were only high enough in time to be affected by the last glaciation.

In Africa, instead of 'glacials' and 'interglacials', geologists recognize **pluvials** and **interpluvials** which probably occurred at the same times. Pluvials were periods of very heavy rainfall with the formation of extensive lakes. They are particularly well represented in the Rift Valley of eastern Africa and in the continuation of this structure in the Middle East. In the same areas, apparently at the same time, there occurred great outbursts of volcanic activity. Before he went to Antarctica, Sir Vivian Fuchs served as a geologist in the much warmer climate of East Africa. He has suggested that there was a direct connexion between the extensive volcanic eruptions of the Pleistocene and the glaciations, due to the screening effect of clouds of volcanic dust in the atmosphere. It may be argued that other geological periods had at least as much volcanicity as the Pleistocene without accompanying glaciations, but this is not a matter which can easily be judged.

Perhaps the most interesting of all events in the Pleistocene was the emergence of man as we know him today. The first creature which can truly be called man appeared within the Pleistocene, along with the first true horse and the first true elephant. Before the end of the period he was making the tools which are used as index fossils in the later deposits. Man probably did not evolve in the glaciated areas, but he was able to adapt himself to the harsh conditions of the Pleistocene better than many of the larger mammals which became extinct before the end of the epoch. The earliest known record of our own species comes from Swanscombe in Kent, but there were also other species—uncouth to our eyes—such as Neanderthal Man (see Figure 60). These shambling creatures, with protruding brows and retreating chins, were quite common in Europe during the Pleistocene, but appear to have been ousted by *Homo sapiens* in an early spell of imperialism.

The position of man today—from a geological point of view—will be discussed in the next chapter, but the most noteworthy fact about him in this history is his modernity. If the whole time since life first appeared on the earth were expressed as one year, then the first vertebrates did not evolve until about October 20th, mammals appeared on December 7th and man himself only stepped into the limelight at twelve minutes to midnight on December 31st.

XII

THE PRESENT AND THE FUTURE

This chapter is chiefly concerned with the present state of the world, its natural processes and its inhabitants (especially man). These matters are considered from a geological point of view and are related both to the past and to the future.

From the end of the last Pleistocene glaciation to the present day, our record is largely in the hands of the archaeologist and the historian. During that time there have been minor oscillations in climate, there have been up and down movements of land and sea and a few large animals have become extinct. Also in this moment of time there has been the whole of human history, the rise and fall of many civilizations and the evolution of all the languages, cultures and religions of the world.

It is hardly worth while here to consider the petty detail of these ephemeral events, but it is worth while to consider the natural processes going on at the present day, since it is from the study of these that we have interpreted the past.

UNIFORMITARIANISM

All through this book there have been constant cross-references from the record of the past to what is known to happen at the present day—the way coral reefs grow in tropical seas, the way lavas are erupted in Iceland, the way sediments accumulate in central Asia and so on. It is the film director's method of 'flash-back' in reverse and it is the basis of the reasoning by which we interpret the geological record. This is called the **'Doctrine of Uniformitarianism'** which was mentioned in Chapter I; it is the principle that the present is the key to the past and that we can interpret all past events in the light of processes seen to be going on today.

For many hundreds of years the study of rocks and fossils made virtually no progress because this simple principle was not recognized or accepted. It was first formulated by the Scottish doctor, farmer and natural philosopher James Hutton, who lived in Edinburgh in the second half of the eighteenth century and who expressed his ideas in his *Theory af the Earth* published in 1795. This, however, is a very difficult work to read and it was left to others to propagate Hutton's ideas.

The chief of these was Sir Charles Lyell whose book *Principles of Geology* was the foundation of the systematic study of the subject. Lyell rejected completely the older ideas of 'catastrophism' and maintained that it was not necessary to postulate any processes different or more severe than those operating at the moment. He showed that the whole of the geological record could be explained by the action of wind and rain, heat and frost, running water and all the other natural processes which affect our modern world. The earth's surface has continually been changing; it has been and is still, always being transformed into something different.

These ideas were revolutionary and rather shocking to nineteenth-century thinkers, who regarded the existing natural world as exactly the same as it had been when first divinely created. It is interesting today to recall the heart-searchings, the agonies of doubt and the sanctimonious tirades which went on in mid-Victorian England as a result of the discoveries of geology. The Victorians found their self-satisfaction and self-confidence shattered when the tidy chronology of the Old Testament, with its clear account of the creation of the familiar world in six days, was replaced by an earth evolving through millions of years by the action of commonplace natural processes.

They were even more disturbed when they were told by Darwin and Wallace that the plants and animals, including man himself, had also evolved through the ages and were far from what they had been at the supposed first creation. The evolutionary ideas of Darwin and Wallace cannot be separated from geology, for though they had little support from the palaeontological evidence available at the time, they sprang logically from the teachings of Lyell and were essentially geological in outlook.

Lyell's *Principles of Geology* considerably influenced Charles Darwin, and he carried a copy of it round the world with him on his *Beagle* voyage. From Lyell's teaching on the inorganic world, it was inevitable that Darwin and Wallace should extend the same ideas to the organic world. In fact (though not generally realized) James Hutton himself had outlined the principle of the 'survival of the fittest' in manuscript notes many years before Darwin was born.

Darwin's fundamental conclusion was that organisms have not always been as they are now, but have evolved through millions of years as a result of processes which can be seen operating today. This was uniformitarianism applied to animals and plants. The natural processes in this case were the intrinsic variability of species operated on by natural selection of the 'fittest'. Darwin's theory preceded the discoveries of genetics, which started from the work of the German monk Mendel growing peas in his monastery garden at Altbrünn in what is now Czechoslovakia. Mendel's overlooked discovery of the principles of inheritance and the later discoveries of mutations and genetic change provided a mechanism for the production of variation within species. This made the operation of evolution by natural selection much easier to understand.

The publication of the *Origin of Species* in 1859 also preceded most of the important fossil discoveries which have become the classic examples of evolution. Darwin could not produce much palaeontological evidence to support his arguments. He had to apologize frequently for the inadequacy of the fossil record, but since his time masses of evidence has accumulated in groups as varied as oysters and horses, corals and man, which clearly demonstrate evolution in action.

THE INORGANIC WORLD

When we consider the processes going on in the world today, we must bear in mind its geological setting. So far as Europe is concerned, we are still in the aftermath of the Alpine Orogeny. The mountains are still young and are still in the early stages of being worn down to the inevitable plain. We are also in the aftermath, or perhaps still within, a period of glaciation. There is probably more land above the sea at the moment than there was during most of the geological past. It is also probable that the climate of the world at the moment is unusually cold and the tropical belt is considerably restricted.

We may suppose that in the near geological future the present 'new' mountain ranges such as the Alps will be worn down almost to sea-level and that this will be followed by a gentle but very extensive marine transgression. Only slight changes would be necessary to melt the polar ice-caps and to flood vast areas of lowlands, including most of the world's capitals. It would be interesting to know how man (if he still exists) will adapt himself to these changes.

If we concentrate our attention (as usual) on the British Isles, we see that erosion and deposition are going on as they always have done. Man has interfered with the natural processes a great deal, but he has not been able to stop them. In many places he has speeded up erosion by cutting down forests and breaking up the soil; in others he has slowed it down by building embankments and breakwaters. But he cannot replace on the summit of Snowdon every pebble that is carried down one of its rushing streams; he cannot entirely resist the insidious advance of sand-dunes across the Morayshire country-side shown in Plate 19. The most important geological processes going on in Britain at the moment are the local advances of sea on land and land on sea.

The speed at which the sea can wear away a coastline is most clearly seen in East Anglia where thousands of acres have been lost in historic times. The most spectacular loss was Dunwich in Suffolk, which was virtually capital of the region during its prosperous Middle Ages. The sea slowly ate into the city and the last of the many medieval churches finally fell from the cliff in 1919. A few blocks of stone and some scattered bones from the churchyard are all that remain on the beach. The flourishing mediaeval borough was so reduced that by the time of the Reform Bill in 1832, less than a dozen electors were sending two Members to Parliament.

In places the sea may be advancing on this coast at a rate of six feet a year (as on the section of the Yorkshire coast shown in Plate 21). It has been calculated that Norfolk and Suffolk lose about thirty-six acres a year. If it went on at this rate, the whole of both counties would be washed away in about 60,000 years. This is no time at all geologically speaking, so we may have here a marine transgression in progress. Just as so many times before in our history, the older rocks are being planed off below, and in due course younger ones will be laid unconformably on top.

The erosion of the coast of East Anglia is an example of direct lateral aggression by the sea; there is also the more insidious process of the sinking of the land which was discussed in the last chapter. London, for example, is at the moment sinking slowly but inevitably year by year. It is an ironic and perhaps significant fact that the Bank of England is sinking more rapidly than anywhere else and is expected to be awash in about 5,000 years time.

In other areas the land is gaining on the sea. Thus at the north end of Cardigan Bay in Wales there is a broad flat area of sand-dunes and marshes between Harlech and Criccieth which has recently been a great inlet of the sea (see Figure 62). When Harlech Castle was built in 1286 it is thought to have stood on the sea-shore, and where there is now a railway station is thought then to have been the wharf where supplies were unloaded for the castle. This retreat of the sea must have been a very recent event geologically speaking, for in this same area there is good evidence of the submergence of a great deal of low-lying ground since glacial times. This ties up with the numerous Celtic legends of the lost land of Cantref-y-Gwaelod in Cardigan Bay which was allegedly flooded because of the carelessness of a drunken sluice-keeper. A prominent submarine ridge—Sarn Badrig—runs out to sea for twenty-one miles near Harlech, and may have formed part of the flooded land.

We have to go farther afield to see some of the other geological processes in operation. We have to study the deserts of North Africa or Central Asia in order to interpret our Permian and Triassic rocks (e.g. by comparing the sandstone bedding shown in Plate 13 with the bedding in modern sand-dunes such as that shown in Plate 20). We find limestone being deposited at the present day on the Bahamas Banks in the West Indies. We find Coal Measure conditions in the Everglades of Florida.

Volcanic activity is going on in many parts of the world today and will no doubt continue with varying intensity. It is concentrated in particular regions as it was in the past. Thus in Europe, volcanic activity today is restricted to the Mediterranean area, just as in early Caenozoic times it was concentrated in the areas bordering on the North Atlantic. All round the Pacific there is what has been called the Pacific 'girdle of fire' with scores of active volcanoes, and associated crustal instability. Fault movements causing earthquakes are well known all round the Pacific, for example the San Andreas fault and the San Francisco earthquake which have already been mentioned. Even in a stable area such as Britain, there are occasional slight tremors which can be attributed to faults such as that of the Great Glen.

Example after example could be selected in this way to match the events quoted in previous chapters. The only type of geological event not readily demonstrable in our fleeting present day is an orogeny buckling up the sediments of a geosyncline. But we know now that these are in fact going on, in some form, wherever one plate is converging with another.

The most obvious part of the world to be thought of as in the midst of an orogeny at the moment is the East Indies. This is known to be an extremely unstable region, both from the violence of its volcanoes and earthquakes and from geophysical observations of the state of the crust. Quite modern coral reefs are found hundreds of feet up on some of the islands.

The East Indies consist of a parallel series of deep-sea troughs, separated by ridges which appear as arcs of islands. The concentric arcs can be continued up into Burma where (as in some of the islands themselves) the deep troughs are filled with very thick and often oil-bearing Tertiary strata. Whether the troughs are empty or full seems to have depended simply on the availability of sediment. On either side of the East Indies there are stable regions or plates as we would call them now, which are moving together to produce an orogeny.

There are island arcs in many other parts of the world, for example the West Indies and the Aleutians. Such island arcs may be closely involved in plate tectonic theory (see Figure 4). They seem to be a regular feature all through the stratigraphical record. They are always associated with thick troughs of sediment, volcanism and subduction zones. There were probably island arcs in the Tethys, very similar to those of the East Indies, prior to the Alpine Orogeny. One is in fact shown in the reconstructed geography of early Tertiary times in Figure 51.

THE ORGANIC WORLD

Lastly in this book we must consider the organic world of today and tomorrow. Here again man's actions have been felt both in destruction and in preservation. There is no doubt that the process of extinction of old species and the evolution of new ones is still going on. Many large animals have become extinct in very recent times. The giant ground-sloths of South America appear to have been kept as domestic animals by early man before they finally disappeared.

226

The flightless bird known as the Dodo was well known to Portuguese and Dutch sailors visiting the island of Mauritius and a live specimen is thought to have been exhibited in London in the seventeenth century, but it was finally exterminated by man and his domestic animals in 1681. The Great Auk, which was widely distributed in the North Atlantic (including islands off Scotland) lasted until 1844. The buffalo of western North America was very nearly exterminated by the Indians and the European pioneers, and was only saved from extinction by special protection.

Other species are still comparatively abundant but are, in fact, only lingering remnants of what were once great races. The two living species of elephant—the Indian and the African—are but scanty survivors of a group which in Pleistocene times ranged all over Europe, Asia and America.

The plants too have similar examples. The redwoods are common in Mesozoic and Tertiary rocks all over the northern hemisphere. Today they are restricted (in natural circumstances) to two species in California and one species in a tiny area in central China. The latter was only discovered by chance in 1945. The Ginkgo or Maiden-hair Tree (which Darwin called a 'living fossil') belongs to a primitive group of cone-bearing plants which goes back at least to the Trias, but would be extinct today but for the preservation of the single living species in Chinese and Japanese temple gardens.

If we consider the humbler groups of animals which are much better known in the fossil record, we find further examples of rise and decline. The brachiopods which were so important in Palaeozoic and Mesozoic times are now very restricted in variety and have been ousted from their old habitats. The gastropods or snail family have only quite recently risen to their present dominating position among the shell-bearing invertebrates. Only in the Caenozoic did they begin to show signs of over-specialization and perhaps senility with features such as long spines. It may be that they are just reaching the climax of their career and will soon begin a gradual decline.

To be honest one must admit that every man is chiefly, if not entirely, interested in himself. This is very obvious when one comes to the question of human evolution and the author makes no apology for discussing his own species at greater length than any other organism. The writer of the standard textbook on vertebrate palaeontology drily remarks when he comes to the primates, 'Fossils are few, but the literature is vast.' For this reason it is not possible to

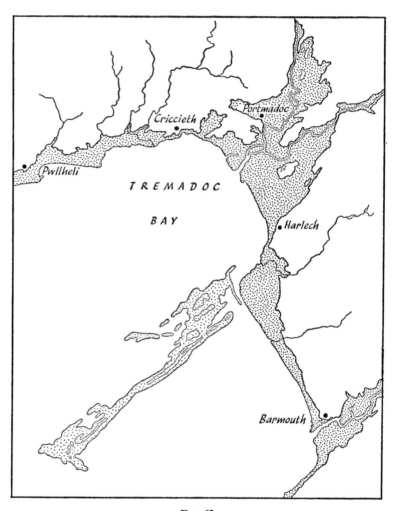

FIG. 62

Map showing recent gains and losses of the sea in the north part of Cardigan Bay. The heavy stipple shows areas which are occupied by recently developed sand-dunes and marshes. The faint stipple indicates a submerged ridge marking the former extension of land into Cardigan Bay (re-drawn from *Coastlines of England and Wales* by kind permission of Professor J. A. Steers).

more than toy with the subject here, but it is worth while to consider man in the light of the palaeontological record.

No intelligent man can today possibly deny the fact of human evolution, but it is not proposed here to dig over yet once more the gradually accumulating evidence. The author prefers to consider some of the more interesting implications of human evolution, seen from the point of view of a palaeontologist who has in mind the great mass of evidence of the form of evolution in animals ranging from oysters to horses. They are completely personal views, but they seem to be inescapable.

As was pointed out in the last chapter, man appeared very recently in the record, but he has proved more successful than all the vertebrate animals which preceded him. It is not so easy to compare him with ubiquitous organisms such as ants, earth-worms and lichen. We tend to think of man's success chiefly in terms of material and intellectual achievements, but he colonized the earth long before he had made much advance in these fields. Probably no other animal of similar size (and certainly not a single species) has spread and become so abundant in so many different environments. His habitat ranges from the hottest tropical rain-forest to the coldest arctic deserts, from the arid hearts of continents to remote oceanic islands, from the lowest river-side marshes to the high mountain plateaux. Perhaps the most remarkable environment of all to which he has become adapted is that of modern cities, living in boxes, travelling in tunnels and utterly dependent for his food on people he never sees and who may live on the other side of the world.

Considered in terms of anatomical evolution, man is not really so very unusual in anything apart from his brain. His bipedal habit was developed millions of years ago in the dinosaurs and in the birds, his limbs are less specialised than those of the horse or the bat, his skull is much less specialized than that of the elephant or the whale. Even the use of tools is not completely unknown outside mankind.

Within his own family of the primates, man is much closer to the common ancestor in certain characters than are some of the anthropoid apes. Thus his arms are not so long relative to his body as are those of the tree-living apes. This is not because he came down from the trees, as some still suppose, but because he is closer to the ancestors who never went up them.

The two features of human evolution which seem to the author to be most unusual are firstly man's limited degree of speciation and

229

secondly the way in which man is, by his own efforts, transgressing the so-called 'laws' of natural selection.

The problem of speciation centres around the fact that there is only one living species of the genus *Homo*. There have been others in the past, such as *Homo neandertalensis*, but using the criterion of potential interbreeding, there is only one today. Judging from the past record of successive stocks—such as the ammonites or the graptolites—one would have expected the successful establishment of man with his innovation of a reasoning ability, to be immediately followed by rapid radiation into many fields of adaptation. That this has not happened may be explained by the fact that we are still right at the beginning of the story. Just as there may have been a single species of armoured dinosaur which gave rise to all the later multitude of strange forms, so we may belong to the ancestral species of man from which a whole phylum of complex stocks may arise. From the present evidence, however, this seems to be extremely unlikely.

If we consider the present variation in our species, it is quite considerable. The different major races of man would certainly be regarded as distinct sub-species had they been birds, and they would even be distinct species if they belonged to some other animal groups. They would certainly be called separate species if they had only been known as fossils. But we know that all the races of man are capable of interbreeding and of producing fully fertile offspring. They are therefore, by every reasonable definition, a single species.

The different races of man are, to a certain extent, adapted to their particular environments as is every other organism. Thus there is adaptation in the devices for retaining and losing heat from the body. The surface area of skin tends to be larger relative to weight in races which live in hot, wet climates than in those which live in cold, dry ones. The legs and arms of negroes tend to be relatively long, whilst those of eskimoes are relatively short. This is comparable with what occurs in other species; foxes and rabbits have short ears in the arctic, but very long ones in the tropics. Similarly, peoples who live under natural conditions in densely populated areas tend to be small in stature. This presumably results from the selection of those stocks able to survive on a smaller food supply.

The normal process for the evolution of new species seems to be the splitting up of the parent stock into populations living in different areas or in different ways. Birds of a single species become separated

230

by seas and by mountain ranges and adapt themselves to slightly different modes of life. If the geographical or habit barriers persist, then the populations come to differ from one another and become what zoologists would call geographical sub-species, though they would still interbreed if brought together again. Eventually the separation would result in distinct populations which would not interbreed and we would then have separate species.

This process seems to have begun in man, but to have been halted by his own efforts. The pygmy was presumably becoming better adapted to his luxuriant tropical forest, the bedouin more able to withstand the cruelty of the desert and the South-Sea islander more at home in his semi-aquatic life of fishing. These were the very early beginnings of speciation. The different races were never likely to meet or interbreed and with the passage of hundreds of thousands of years would have become more differentiated in physical features. It should be borne in mind that the differences already visible must have arisen in the few thousand years since the races moved into their present habitats. Even the senselessness of racial prejudice may be interpreted scientifically as a facet of the trend towards non-interbreeding between the more different races.

So we may presume that within a few thousands or millions of years there would have been many different species of *Homo* living on the earth. It would not be possible to say that one species was 'better' or 'more advanced' than another, they would just have been different and living in different ways. The taxonomists of the future would probably have a problem in deciding which species should retain the old name of *Homo sapiens*.

The coming of modern transport and social habits in the past few hundred years has completely reversed the natural process of speciation. There is now no likelihood at all of the different races becoming more and more distinct. A few have and will die out like the Tasmanian aborigines, but the majority are inevitably converging again towards a single variety. In South America it has been estimated that there are now as many of 'mixed blood' as there are 'pure' Indians, negroes and whites. In spite of all the ignorant resistance of the racialists in every country, the different geographical variants have been brought together again. They had not differentiated far enough to become distinct species, so they must inevitably become a single interbreeding population. This is not a matter for the will-o'-the-wisps of morality and politics, it is a simple biological trend.

THE PRESENT AND THE FUTURE

The second anomaly referred to earlier was the way in which man has gone against the 'laws' of natural selection. This has already come into the matter of speciation, but it is seen most clearly in the problems connected with man's protection of his weaker fellows against the harsh law of the 'survival of the fittest'. Again we come to problems of morality and politics, but the underlying processes, viewed dispassionately, are purely palaeontological.

The cleft palate and hare lip are unfortunate defects in human babies which are strongly inheritable. Modern surgery and other methods are now making it possible for such children to develop a 'normal' appearance, and therefore make it more likely that they will marry and reproduce themselves. There are already signs that these disfigurements are becoming more common with succeeding generations. This may be regarded as a purely ethical question involving the fundamental value of human life and happiness, but it is also depressingly intriguing to a palaeontologist with its implications for future human evolution. The late Bishop of Birmingham used to speak of the survival and reproduction of mental defectives as a cloud on the horizon which contained an ominous threat for the future. Coupled with such problems is the tendency of the less educated, the less healthy and the less capable members of the species to reproduce themselves at a greater rate than those at the other end of the scale. Deficiencies of environment can be overcome, but those of genetic make-up cannot and must inevitably weaken the human stock.

We are by our own efforts cancelling out the process of 'survival of the fittest' as it applies to ourselves. It is becoming so that the fittest have no more chance of survival than the less fit, and may in fact drop behind in the matter of reproduction. This can be seen in the course of a couple of generations. The self-reliant, adaptable street-urchins and manual workers of the beginning of this century are not, by all accounts, to be compared in terms of fitness for survival with their Welfare State equivalents of today. This is not a judgement; the present author would be the last to favour a return to the harsh, more selfish world of seventy years ago, but it is a contradiction of so-called natural laws. A comparable physiological trend is seen in the fact that in the past few decades there has arisen in civilized societies a much higher proportion of mothers who cannot produce enough milk to feed their babies themselves. The babies do not now starve and the trend is continued.

We may wonder, a little fancifully perhaps, what is going to happen next. In our more urbanized societies man is presumably becoming less adapted to natural environments. The American with his central heating and air conditioning already appears to be more sensitive to slight changes in outdoor temperature than does the visitor from the frigid homes of Britain. Similarly the state of the sidewalks away from the city centres in the U.S.A. already suggests that car-conveyed humans will soon (geologically speaking) lose the use of their legs.

One important factor in human evolution, as in all evolution, is the usually conservative influence of what Darwin called 'sexual selection'. If it were true that 'gentlemen prefer blondes' then we might expect (probably quite wrongly) that the world would be dominated by fair-headed people in a few generations. On the other hand, unusual variants of the human species are not likely to find a mate and to reproduce themselves. A human male mutant with a head two feet wide in a Wellsian manner, might have exceptional intellectual powers, but it would be most unlikely that he would ever marry and continue his mutation.

The most remarkable demonstrations of rapid evolution have been those in which man has taken natural selection into his own hands in the breeding of domestic animals and plants. We have the fantastic and often functionless variety of such species as the dog and the pigeon. This has never been tried on man himself (though Hitler and some earlier tyrants hinted at it), and one hopes that it never will. There are interesting potentialities, however, in human habits of mate selection which never arise in other species. Thus musicians tend to marry musicians, mathematicians tend to marry mathematicians and geologists to marry geologists (though this last character, at least, is probably not inheritable).

But these things are all, to varying degrees, hypothetical and based on assumptions of unequal validity. If we try to pin down what exactly is happening in the evolution of *Homo sapiens*, we find that we have very little to go on. In terms of purely physical evolution, we do not seem to have changed much since the first emergence of the species. We can make isolated observations such as the fact that the average height of Englishmen has increased by about two inches in the last two or three hundred years. We can repeat once more, rather smugly, the old story of the mediaeval suits of armour which are far too small for the men of today. But these are reflections of changes in nutrition and in general standards of living rather than evolution. We do not compare so well with the fine specimens of

233

Crô-Magnon Man who was among the first of our species. We may have improved slightly in our resistance to certain diseases by the elimination of the more susceptible, but most of the perceptible physical changes have been degenerative. The toes of modern man are clearly losing their function, as are his teeth, which often do not serve for half a lifetime. Our hair is probably disappearing rapidly and may be gone in a few hundreds of thousands of years, as will our sense of smell. But these are very minor points and man has largely overcome his physical environment. It is not likely that natural selection in the sense of organisms battling with their environment will much further influence our species.

If we turn to the evolution of the intelligence, here too it is doubtful if we have changed very much since the late Pleistocene. Modern man may know more than his ancestors, but there is no evidence whatever to prove that on the average he is any more intelligent than the Ancient Greeks.

There remains a third type of evolution—that of man's society—and this is the only field in which any appreciable progress has been made in the last twenty thousand years. By learning to organize ourselves, to divide the different kinds of labour between different people, to co-operate on major tasks, to educate the young and generally to be social rather than anti-social, we mastered the earth's surface and all its inhabitants and are no longer much affected by our environment. We have minor local setbacks from time to time—from earthquakes and droughts and mosquitoes—but our evolution is not likely to be altered by marine transgressions or the other geological changes we have found in the past. We may even—as Olaf Stapledon suggested in his *Last and First Men*—eventually be able to overcome even astronomical changes, such as the expected future heating-up of the sun.

We may degenerate through our rejection of natural selection, we may be displaced by some better-equipped organism (though there is nothing on earth at the moment which faintly approaches this category). Most likely of all, we may destroy ourselves by our own efforts, but on the whole it looks very probable that man and his products will be the zone fossils of the strata accumulating for a very long time to come.

> *We do not know very much of the future*
> *Except that from generation to generation*
> *The same things happen again and again.*

APPENDIX

THE STUDY OF GEOLOGY

This final section is intended as a guide for those who wish to go further in the study of geology, either as a possible profession or as a fascinating hobby.

The geologist is a fortunate man who combines mental and physical exercise in an open-air occupation that always has something in it of the excitement of discovery. There can be few indeed whose work is essentially intellectual and yet who spend so much of their time out of doors.

SOCIETIES AND PUBLICATIONS

Britain has, for its size, a large number of societies concerned with all branches of geology. The senior society, and the oldest geological society in the world, is the **Geological Society of London**, with splendid apartments in the east wing of Burlington House facing on to Piccadilly. It was founded in 1808 and holds regular meetings for the reading and discussion of scientific papers. It has a huge library of geological publications (including maps) for the use of fellows, and publishes a quarterly journal.

This is very much the professional society and comparatively few amateurs apply to be elected fellows. The annual subscription at present is £10.50.

The **Geologists' Association**, founded in 1858, has always been considered the chief organization for amateur geologists in England, though nowadays—with the increasing importance of science in society—there is probably a majority of professional members. It has the interesting and desirable custom of alternating between amateur and professional geologists in its succession of presidents.

The 'G.A.' also regularly holds meetings at Burlington House as the guests of the Geological Society, and it issues a quarterly publication. It has a large library which is combined with the science library of University College, London, in Gower Street. The association is less inhibited than the senior society in its activities, and an important part of its programme is the holding of numerous field meetings and demonstrations. The former range from half-day trips near London to two or more weeks in places as far away as Bulgaria and Cyprus. The annual subscription at present is £4. The Geologists' Association has some local groups outside London but its activities are mainly concentrated in the southern part of Britain.

There are also some very notable local geological societies, for example the **Geological Society of Glasgow**, the **Liverpool and Manchester Geological Society,** the **Yorkshire Geological Society** and the **Royal Geological Society of Cornwall**; also many naturalists' societies such as the **London Natural History Society** and the **Cotteswold Naturalists' Field Club** have strong geological interests.

Geologists played a large part in the formation of the **British Association for the Advancement of Science** and geology is third on its list of sections. The 'British Association' meets for a week every year at a different city and the local geologists always arrange excursions to demonstrate the nearby rocks.

There are also more specialist societies. There is the **Palaeontographical Society** which owes its formation to the discovery of large numbers of fossils in the London Clay during the excavations connected with the building of the Highgate Archway. The main function of this society is the publication of monographs of British fossils. It is the oldest palaeontological society in the world and had the unique distinction among scientific societies (until quite recently) of not having increased its subscription in over a hundred years. Of much more recent origin is the **Palaeontological Association** which holds meetings in various parts of the country and publishes shorter papers on fossils. The **Mineralogical Society** concerns itself in the same way with minerals, though the complications of modern techniques and apparatuses make it an unknown country to many geologists.

All the above societies have their own publications, which deal with all branches of geology, mostly in the form of scientific papers recording new observations or new theories. There is also the **Geological Magazine,** published bi-monthly at Cambridge, which includes short papers, reviews and correspondence. There are also

THE STUDY OF GEOLOGY

innumerable foreign and commonwealth publications of the same type and the publications of all the geological surveys and many museums. So far as general textbooks on geology are concerned, a short annotated list is given at the end of this Appendix. These are all fully intelligible to the ordinary reader and are easily obtained in Britain.

The excellent libraries which the Geological Society and Geologists' Association have available for members have already been mentioned. There are also very large geological libraries at the Natural History Museum and the Geological Museum in South Kensington, which are open to the general public, but from these, of course, books cannot be taken away.

GEOLOGY AS A PROFESSION

To take up geology as a profession, a university degree is essential. Practically all but the newest universities in the British Isles and several of the new Polytechnics have geology departments, and at nearly all of these it is possible to take an honours (or 'special') degree in the subject. In some departments, specialization in one branch of geology (e.g. palaeontology or mining geology) is required at degree level, but at most one studies all aspects equally. Entrance regulations differ considerably from university to university, but generally speaking a good scientific background is desirable.

The vast majority of geological posts are abroad—especially in the less-developed parts of the world where primary exploration is still going on. A necessary qualification therefore for a potential geologist is adaptability to life under varied and often difficult conditions. Apart from occasional posts in universities, museums and the home geological survey (for all of which there is considerable competition) there are few opportunities for geologists in Britain. As a result of this, there are few posts available for girls in geology, though the few who do take it up always seem to find suitable posts for the few years before they marry male geologists.

At the moment there is still a considerable demand for geologists in most parts of the world, and this is likely to continue, though in places the demand is closely controlled by the market price of a particular mineral. Both the U.S.A. and the U.S.S.R. employ far more geologists in proportion to the size of their populations than we do in Britain. In Russia the subject is even considered sufficiently important to have its own government ministry. Communist China is at present training almost incredible numbers of geology students in an urgent attempt to find the raw materials for her growing industries. A recent film from China showed a party of literally scores of geologists (of all specialities) descending on an area near the Tibetan frontier in search of iron-ore. In another scene a party of women geologists, mounted on camels, were prospecting an area in Manchuria. In Canada on the other hand, the geologists are not so numerous, but are more mechanized, and geology has become a matter of helicopters and motor-boats.

GEOLOGY AS A HOBBY

For those who do not want to become professional geologists, the subject still has endless possibilities as a hobby and as a subject in which almost anyone can make a useful and original contribution. Some of the foremost world authorities on certain aspects of the subject are amateurs.

All one needs in the way of equipment is a hammer, a cold chisel, a map, a willingness to walk and to get wet, and a good supply of patience. There is so much to be found out about the past—even in a well-studied area like Britain—that it is a pity not to make one's contribution. Almost everyone can make original observations about his local rocks and fossils, though he should be very careful indeed before committing such discoveries to print.

A fascinating occupation for anyone living in a suitable area (and almost every area is suitable) is making one's own geological map. Starting with the ordinary large-scale Ordnance Survey topographical map (the scale of six inches to one mile is most suitable) one slowly builds up a picture of the rocks of the neighbourhood. The map grows before your eyes as you mark in the outcrops of different rock types. The book by Himus and Sweeting in the list below gives clear instructions on how to go about it. In many areas—especially in lowland Britain—there are few exposures of solid rock, but it is surprising how much can be learned by studying the soil and slight differences in the slope of the land. A hand auger can often be used successfully in such areas. Many amateur geologists become great experts on the glacial deposits—the boulder clays, sands and gravels—which others regard as merely the rubbish on the surface. These have a fascination all of their own and are no more worthy of disdain than any other part of the stratigraphical column.

The experts at the Natural History Museum and the Geological Museum are usually willing (given a little time) to identify specimens, though it is most important that every specimen should be clearly labelled with its locality and (if possible) the exact stratum from which it was taken. The author does not wish to make himself too unpopular with his friends in these excellent institutions and so would urge the reader not to go there with boxes of miscellaneous unlabelled junk merely out of idle curiosity. Well-preserved specimens are always interesting to the specialist and so are even poorly preserved one if they appear to be unusual or come from supposedly unfossiliferous strata. If a specimen is particularly rare or interesting, the specialist may invite the finder to present it to one of the national collections, where it is clearly of much more use than in a private collection, but there is no compulsion about this.

Great care should always be taken to keep the labels with the specimens and preferably to give them catalogue numbers as well. A specimen without its label is virtually useless and nothing is so frustrating as to sort over an old abandoned collection full of wonderful things which are completely useless because no one knows where they came from. A collection of rocks or fossils very soon gets out of hand and unmanageable, and will try the patience of the most tolerant wife or mother. It is best only to keep what is really worth while and to concentrate one's attention on a particular area, a particular formation or a particular group of fossils. Mere 'stamp-collecting' is both unscientific and ultimately unsatisfying.

RECOMMENDED READING

H. H. Read and J. V. Watson, *Introduction to Geology* (Macmillan). A very readable introductory text-book.

A. Holmes, *Principles of Physical Geology* (Nelson). The great classic of geology that is remarkable for its breadth of cover and clarity of exposition.

G. W. Himus and G. S. Sweeting, *The Elements of Field Geology* (University Tutorial Press). A suitable book for the beginner, full of information and suggestions about the strictly practical side of geology.

H. H. Read, *Rutley's Elements of Mineralogy* (Murby). The standard elementary textbook, several times revised and brought up to date.

A. Morley Davies, *An Introduction to Palaeontology* (Murby). The most readable of the introductory textbooks on palaeontology.

I. G. Gass, P. J. Smith and R. C. L. Wilson (editors), *Understanding the Earth* (Artemis Press). This and the other publications of the Open University let a breath of fresh air into the subject. They are particularly good (at times idiosyncratically so) on the geophysical side of the subject.

A. Hallam, *A revolution in the earth sciences* (Oxford Univ. Press). A very readable account of the growth of ideas about the 'new geology'.

The beginner in geology should also know and make use of the publications of the Institute of Geological Sciences. These include a series of eighteen regional handbooks, each dealing with a natural region of Great Britain. They are of varying quality, but are always extremely useful. There are also more detailed local memoirs relating to the published one inch to one mile geological maps.

There is also a very useful series of simple guides, published by the Geologists' Association, with suggested routes to be followed in particular areas; and a useful and readable series of books published by David & Charles entitled *Geology explained in* . . . . So far volumes have been published on: Severn Vale and Cotswolds (by W. Dreghorn), South Wales (by T. R. Owen), North Wales (by J. Challinor and D. E. B. Bates), South and East Devon (by J. W. Perkins), Forest of Dean and the Wye Valley (by W. Dreghorn), Dartmoor and the Tamar Valley (by J. W. Perkins).

One of the most useful tools for anyone interested in geology is the two-sheet geological map produced by the Institute of Geological Sciences at a scale of ten miles to the inch.

MORE ADVANCED READING

STRATIGRAPHY

B. Kummel, *History of the Earth* (Freeman). A world-wide stratigraphy, though with an American bias.

M. Gignoux, *Géologie stratigraphique* (Masson). A world-wide stratigraphy, though with a Gallic bias. Written in very readable French, but translated into less readable American.

G. M. Bennison and A. E. Wright, *The geological history of the British Isles* (Arnold). A standard source-book of information about the details of the British record.

D. V. Ager, *The nature of the stratigraphical record* (Macmillan). An idiosyncratic and somewhat light-hearted discussion of the record, though with a serious purpose.

PALAEONTOLOGY

D. M. Raup and S. M. Stanley, *Principles of Palaeontology* (Freeman). The best of the palaeontological text-books that is something more than a telephone directory of fossil names.

D. V. Ager, *Principles of Palaeoecology* (McGraw-Hill). An attempt to approach fossils as animals and plants that once lived and breathed, fed and bred, moved and died, rather than as dusty specimens in drawers.

R. C. Moore (editor), *Treatise on Invertebrate Palaeontology* (Univ. of Kansas Press). A vast work with an army of contributors in x volumes, but always unfinished. Whatever you are interested in is likely to be there somewhere.

MINERALOGY, PETROLOGY, SEDIMENTOLOGY

F. H. Hatch, A. K. Wells and M. K. Wells, *Petrology of the Igneous Rocks* (Murby). An excellent general textbook on igneous rocks, including plutonic types.

F. H. Hatch, R. H. Rastall and M. Black, *Petrology of the Sedimentary Rocks* (Murby). The standard British work on the sediments.

R. C. Selley, *Ancient sedimentary environments* (Chapman & Hall). A compact and lively book discussing a limited number of 'case histories'.

The above is but a tiny selection of a mountain of excellent standard textbooks on different aspects of the subject, including some good (but very expensive) ones published in the U.S.A. and France, but the best way to learn geology is not in books but in the field.

Go and look at the rocks for yourself, but in these days of conservation, please go easy on the hammer!

GLOSSARY OF TERMS

Acid: applied to rocks having a high percentage of silica.

Agglomerate: volcanic rock consisting of large fragments embedded in ash.

Algae: the simplest types of plants including seaweeds and the green growth seen on ponds.

Ammonite: extinct group of coiled cephalopod molluscs with complex partitions between their chambers.

Ammonoids: applied to the group of extinct cephalopods which includes both the Palaeozoic goniatites and the Mesozoic ammonites.

Amphibians: cold-blooded four-footed animals which have gills in youth and lungs later (e.g. frog).

Anthracite: hard, black, shining coal with a low gas content.

Anticline: an arch-like fold in the rocks, with the beds dipping in opposite directions on the two sides.

Arthropods: members of the invertebrate animal phylum characterized by jointed legs (e.g. crabs, insects).

Ash: fine-grained material ejected from a volcano.

Bacteria: minute organisms responsible for decay and many other processes.

Basalt: a fine-grained, basic, igneous rock; the commonest form of volcanic lava.

Basic: applied to rocks having a low percentage of silica and usually rich in iron and magnesium.

Basin: applied to a basin-shaped feature which may be either structural, with the rocks dipping inwards, or purely topographical.

Bauxite: deposit chiefly consisting of aluminium hydroxides formed in tropical or subtropical climates.

Belemnites: extinct cephalopods, distantly related to the squids and

242

cuttle-fish, with a calcareous rod or guard which probably acted as a counterweight.

Benioff zone: an observed line of earthquake epicentres dipping inwards and downwards from the edge of continents.

Bivalves: now used generally for bivalve molluscs with the valves arranged in a left and right position (e.g. cockle); = lamellibranch or pelecypod.

Boreal: northern, often applied to postulated northern seas at various times in the geological past.

Boss: large rounded mass of igneous rock intruded into stratified rocks.

Boulder Clay: deposit of glacial origin consisting of large unsorted blocks in a fine-grained ground-mass (= till).

Brachiopods: members of a phylum of simple, marine, bivalve invertebrates with the valves arranged in a dorsal and ventral position.

Breccia: sedimentary rock consisting largely of angular fragments.

Bryozoans: minute colonial animals, usually encrusting and sometimes called the 'sea mats'; = polyzoans.

Cephalopods: molluscs having the head surrounded by tentacles and commonly having a univalve, straight or coiled shell, divided internally by partitions (e.g. nautilus).

Chert: compact rock of almost pure silica, usually in nodular form.

Cirque: steep-sided, rounded hollow in a mountain, produced by glacial erosion.

Cleavage: a tendency to split along parallel planes, usually developed in a rock as a result of pressure.

Cone-sheets: sheets of intrusive rock, circular in outline at the surface and inclined inwards towards a common point.

Conglomerate: sedimentary rock consisting largely of rounded pebbles cemented together.

Continental Drift: the hypothesis that the continents may have moved relative to one another in the geological past.

Crinoids: plant-like echinoderms, usually fixed to the sea-floor by a stalk and having a cup with branching arms for feeding.

Crustaceans: arthropods with a rigid, outer skeleton, such as crabs and lobsters.

Current Bedding: *see* False Bedding.

Cuvette: a basin in which sediments accumulated, usually in a mountainous region.

243

Cycle of Sedimentation: a regular and repeated sequence of sedimentary rocks.

Derived Fossils: fossils which have been removed by erosion from an earlier rock and incorporated in a later one.

Detritus: the broken-down material produced by the weathering of rocks.

Diachronism: applied to rock types or fossil assemblages which cut across time planes (i.e. they are not of the same age everywhere).

Dinosaurs: a large and varied group of extinct land reptiles, often of large size.

Dip: the slope of a bed of rock relative to the horizontal.

Dome: an upfolded area from which the rocks dip outwards in all directions.

Dreikanter: a pebble shaped by sand-blasting on a desert surface.

Drift: general term for glacial deposits.

Drumlin: an elongated hillock of glacial material.

Dyke: a sheet of intrusive igneous rock that cuts across earlier bedding or structures.

Echinoderms: members of a phylum of invertebrate marine animals characterized by five-fold symmetry and skeletons of crystalline plates of calcium carbonate (e.g. starfish).

Echinoids: echinoderms with a box-like skeleton (= 'sea-urchins').

Era: a major division of geological time.

Erratic: a loose rock different from that of the neighbourhood and transported there by ice or other process.

Escarpment: *see* Scarp.

Evaporite: deposit produced by evaporation (e.g. rock-salt).

Extrusive (extrusion): applied to igneous rocks which emerge at the surface.

False Bedding: bedding in a sedimentary rock which was deposited at a steep angle due to current action, etc. (= current bedding).

Fault: a fracture in the rocks of any dimension.

Felspar: one of the most abundant rock-forming minerals, a complex series of silicates.

Flags (flagstones): a rock that splits readily into slabs, usually sandstone.

Flint: a form of chert characterized by a shell-like fracture; found particularly in the Chalk.

Fold: a bend in a rock of any form.

GLOSSARY OF TERMS

Forams: members of a group of unicellular animals, normally very small and usually with a calcareous skeleton.

Fossil: the remains of any animal or plant in a rock, or a mark made by such an organism.

Fusulinids: a group of exceptionally large spindle-shaped forams characteristic of late Palaeozoic rocks.

Gastropods: univalve molluscs in which the shell is usually spiral and without partitions (e.g. snail, whelk).

Geology: the study of the earth.

Geosyncline: a great elongated downfold in which great thicknesses of sediment accumulate over a long period of time.

Glaciation: a major advance of ice-sheets over a large part of the earth's surface.

Gneiss: a coarsely-banded metamorphic rock.

Goniatites: Palaeozoic ammonoids in which the partitions between the chambers are sharply bent.

Graded Bedding: a phenomenon of sedimentary rocks in which the constituents of individual beds are sorted with the coarsest material at the bottom.

Granite: a coarse-grained, acid, igneous rock with abundant quartz and felspar.

Granitization: the process of alteration of other rocks into granite without actual melting.

Granulite: a metamorphic rock consisting of even-sized grains, and probably resulting from the alteration of a sandstone.

Graptolites: a group of extinct colonial organisms, rather plant-like in appearance, now thought to be distantly related to the vertebrates.

Grit: a sandstone in which the individual grains are angular (also used in the Cotswolds for rubbly limestones).

Hanging Valley: a tributary valley which terminates high above the floor of the main valley due to the deeper erosion of the latter.

Historical Geology: *see* Stratigraphy.

Igneous Rocks: strictly speaking, rocks which have been molten at some time in their history.

Inlier: an outcrop of older rocks completely surrounded by younger ones.

Interglacial: the period of warmer climate between two glaciations.

Interpluvial: the period of dryer climate between two pluvials.

Intrusive (intrusion): applied to igneous rocks which have been emplaced below the surface.

Klippe (pl. klippen): an isolated fragment of a nappe or similar structure left behind by erosion.

Lamellibranchs: bivalve molluscs with the valves arranged in a left and right position (e.g. cockle).

Lava: molten volcanic rock extruded at the surface.

Limestone: a sedimentary rock largely composed of calcium carbonate.

Loess: a fine-grained wind-blown dust deposit.

Magnetic reversal: the tendency of the earth's magnetic field to 'reverse' periodically, so that the north pole becomes the south and vice versa.

Marble: a metamorphosed, recrystallized limestone.

Marl: a vague term in geology for fine-grained clay rocks usually (but not always) with a high percentage of lime.

Metamorphic (metamorphism): applied to rocks which have been considerably altered by heat, pressure, etc.

Mineral: a naturally-occurring chemical element or compound in crystalline form.

Molasse: usually applied to the coarse sediments derived from a recently-elevated mountain range.

Molluscs: members of a phylum of invertebrates, usually with shells, including gastropods, cephalopods and lamellibranchs.

Moraine: an accumulation of detritus at the margins of glaciers or ice-sheets.

Nappe: a large recumbent fold or thrust mass of mountain-range proportions.

Nautiloids: cephalopods with straight or coiled shells and simple partitions, including the living genus *Nautilus*.

Neptunic: sedimentary.

Nummulites: large coin-like forams of the early Tertiary.

Obduction: the process of carrying oceanic material upwards into newly-formed mountains rather than downwards as subduction (q.v.).

Oil Shales: shales impregnated with natural oil.

Oolite: a limestone composed of small spheres like fish eggs.

Ophiolite suite: the common association of pillow lavas, serpentines and radiolarian chert (q.v.) which is taken to represent oceanic crust.

Orogeny: a major disturbance or mountain-building movement in the earth's crust.

246

GLOSSARY OF TERMS

Outcrop: the area where a particular rock formation comes to the surface.

Outlier: an outcrop of younger rocks completely surrounded by older ones.

Palaeobotany: the study of fossil plants.

Palaeogeography (palaeogeographical maps): the reconstructed geography of the geological past.

Palaeontology: the study of past life.

Palaeozoology: the study of fossil animals.

Period: a major division of an Era.

Petrology: the study of rocks.

Pillow Lava: a lava which formed as a series of pillow-like masses, probably under water.

Pipe (volcanic): the tube leading to a volcano, sometimes filled with solidified material.

Plate: slabs of the earth's surface layers (or lithosphere), of which there are six major plates and several minor ones; they are created at ocean rises and destroyed in Benioff zones (q.v.).

Plutonic: applied to rocks which have formed at great depths below the surface.

Pluvial: a period of persistent heavy rainfall, probably corresponding to a glaciation, but occurring at lower latitudes.

Polar wandering: the hypothesis that the poles may have moved in relation to the continental masses.

Polyzoa: a phylum of minute colonial animals, sometimes called the 'sea-mats' (= Bryozoa).

Pterodactyls: a group of extinct Mesozoic flying reptiles.

Quartz: one of the main rock-forming minerals, composed of pure silica.

Quartzite: a hard sedimentary rock composed essentially of quartz sand cemented by silica.

Radiolarian chert: a siliceous sedimentary rock largely composed of the remains of the minute marine organisms known as radiolaria.

Raised Beaches: platforms bearing beach sand and shingle above the present high-tide mark and formed during some past period of relatively high sea level.

Red Beds: a general term for red sandstones, marls, etc. which appear to characterize arid periods in the past.

Regional Metamorphism: the alteration of rocks over a very large area due to some major geological process.

Rift Valley: a major topographical feature produced by the dropping down of a long tract of country between two parallel faults.

Ring-dyke: a more or less circular dyke, usually thought to have been produced by the subsidence of a large cylindrical mass of rock.

Ripple-marks: an undulating surface produced in a sediment by shallow water action.

Rôches Moutonées: small rocks which have been smoothed and 'plucked' by the passage of ice.

Sandstone: sedimentary rock composed of cemented sand-grains, usually quartz.

Scarp: a steep rise in the ground produced either by the outcrop of a resistant rock or by the line of a fault.

Schist: a finely-layered metamorphic rock which splits easily.

Sea-urchin: *see* Echinoid.

Seat-Earth: the bed immediately below a coal seam, composed of clay or sand.

Sedimentary Rocks: rocks formed by the accumulation of sediment derived from the breakdown of earlier rocks, by chemical precipitation or by organic activity.

Serpentine: an ultrabasic rock, characteristic green in colour, derived from the peridotite layer below the ocean floor.

Shale: a clay rock which splits easily parallel to the bedding.

Shelf Sea: shallow seas on the continental shelves.

Shield: a region of very ancient rocks forming the core of a continent.

Silica: the chemical compound of oxygen and silicon which are the two commonest elements in the earth's crust.

Sill: a sheet of intrusive igneous rock lying parallel to the bedding of the rock that is intruded.

Slate: a metamorphosed clay rock with a pronounced cleavage along which it readily splits.

Stratigraphy: strictly the study of the stratified rocks or strata; generally applied to the study of geology as an historical record (= historical geology).

Stratum (pl. strata): a rock layer or bed.

Strike: a line on an inclined stratum at right angles to the dip, along which it would outcrop on a level surface.

Structural Geology: *see* Tectonics.

GLOSSARY OF TERMS

Subduction: the process by which ocean-floor material is carried downwards under a continent via a Benioff zone (q.v.).

Sun-cracks: cracks in a sedimentary rock produced by drying; often filled with later sediment.

Syncline: a trough-like fold in the rocks, with the beds dipping inwards on either side.

System: the rocks which accumulated during a Period.

Tectonics: the phenomena associated with rock deformation and rock structures generally; the study of these phenomena (= structural geology).

Thrust: a fault inclined at less than 45° from the horizontal along which one mass of has been pushed over another.

Till: *see* Boulder Clay.

Transform fault: shear faults terminating abruptly, along which plate movements stop or change direction.

Trap: old name for a lava flow.

Trap Topography: the step-like landscape produced by the weathering of a series of lava flows.

Trilobites: an extinct group of arthropods, possibly related to the crustaceans, with a trilobed dorsal skeleton.

Unconformity: a surface separating two sets of rocks, older and younger, and representing a break in the stratigraphical record.

Uniformitarianism: the doctrine that the past geological record can be interpreted by reference to present-day phenomena and processes.

Vein: a thin and usually irregular igneous intrusion.

Vertebrates: animals with backbones.

Volcanic: pertaining to volcanoes or any rocks associated with volcanic activity at or below the surface.

INDEX

Entries in this index refer only to people, places and major divisions of geological time. Technical terms will be found in the Glossary. Numbers in italics refer to figures.

INDEX

INDEX